I0393598

ROCK CRUSHERS INVENTIONS

S.I. FISHGAL

I TRANSMISSION ROCK CRUSHERS

I.1. DOUBLE-ACTING TOGGLE JAW CRUSHERS

BACKGROUND OF THE INVENTION

This invention relates to overhead eccentric jaw crushers. Such known crushers (e.g. manufactured by Kennedy Van Saun, Telsmith, Iowa Manufacturing Co. and others) consist of a stationary jaw and a swing one mounted directly on the eccentric shaft to receives a downward-forward motion. The lower end of the swing jaw is held in position against the toggle by a tension rod.

The objective of the present invention is to elevate the efficiency of the crushers and decrease its requirements to the strength, balancing, mechanical drive and foundation of the construction.

SUMMARY OF THE INVENTION

The above objective is achieved by providing lower parts of jaws with arc extensions forming a secondary crushing outlet in the converging passage between them and interacting with each other as a hammer and. anvil. By this, the vertical component of the jaw movement and the falling down mass of the jaws are also utilized, whereas in the known crushers only the horizontal component of the movement is used. Thus, the present crusher greatly increases the working displacement and crushing area, the mechanical drive working uniformly, with the less horsepower, dynamic load and mass.

The upper part of the stationary jaw can be also provided with an arc extension wrapped around the interacting eccentric hinge of the opposite swing jaw and forming a primary crushing inlet in the converging passage between the arc and hinge. This elevates above effects still more, since the lifting movement of the jaw is also utilized. In order to facilitate the throughput in the outlet, the swing jaw can be provided with transverse tooth elements working as a scraper conveyor.

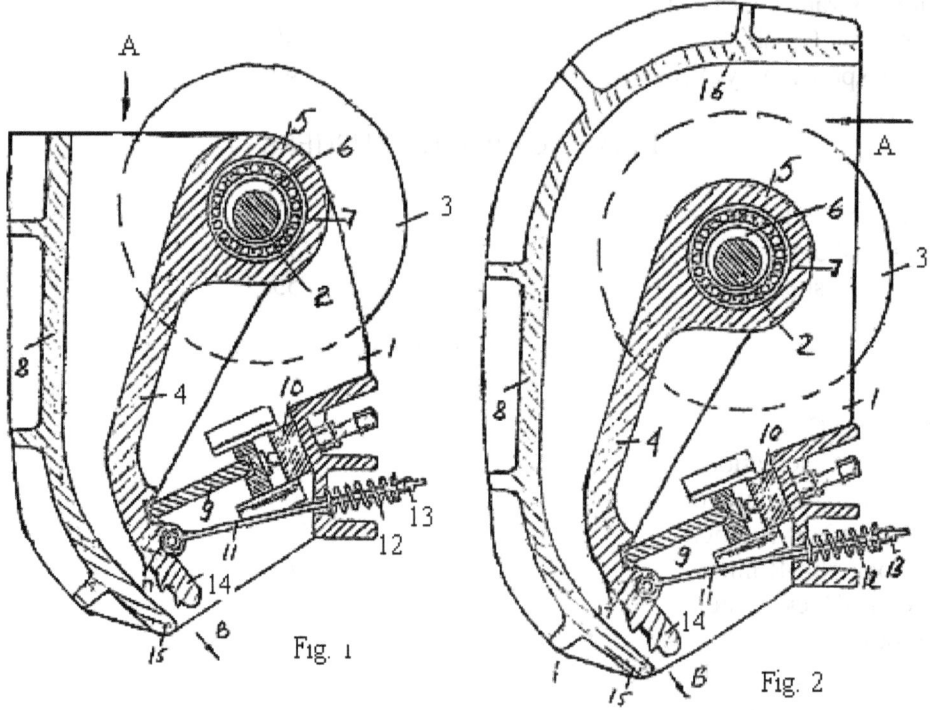

Fig. 1   Fig. 2

IN THE DRAWINGS

Fig, 1 is a cross-sectional view of the present overhead-eccentric jaw crusher with anvil-hammer arc-jaw extensions, the first embodiment.

Fig. 2 is the same as above, with an arcuated jaw top, the second embodiment.

DESCRIPTION OF THE EMBODIMENTS

The crusher of the present invention consists of a frame 1 on which an eccentric shaft 2 with a mechanical drive 3 is mounted in bearing supports (not shown). A swing jaw 4 is mounted with its hinge 5 on the eccentric 6 of the shaft 2 by bearing 7. A stationary jaw 8 is fixed directly to the frame 1.

The lower part of the jaw 4 is releasable and fixed in position by a toggle arrangement 9 serving as a safety mechanism and provided with a series of shims 10 to adjust the position of the jaws 4 and vary the discharge opening between the jaws. The toggle arrangement 9 is held in place by tension rods 11 mounted on the frame 1, provided with springs 12 and threaded ends carrying nuts 13 and extending through suitable holes in the frame 1. The lower parts of the jaws 4 and 8 have arc extensions 14 and 15 forming a secondary crushing outlet in the converging passage between them and interacting with each other as a hammer and anvil.

The upper part of the jaw 8 can also be provided with an arc extension 16 wrapped around the interacting eccentric hinge 5 and forming a primary crushing inlet in the converging passage between the arc and hinge (Fig. 2).

Rock materials are fed into the crusher as indicated with arrow A, are crushed in the converging passages and between the jaws the horizontal and vertical components of the jaw movement. A peristaltic pumping action enhances throughput of the materials through the crusher. Still more facilitating of the throughput is achieved by transverse tooth elements 11 covering partially or entirely the swing jaw working surface and working as a scraper conveyor (in the drawings, the elements 18 are shown, in way of illustration, but not in a limiting sense, on the hammer extension).

I.2. DOUBLE-ACTING DUAL-ECCENTRIC CRUSHERS

BACKGROUND OF THE INVENTION

This invention relates to overhead eccentric jaw crushers. Such known crushers (e.g. manufactured by Kennedy Van Saun, Telsmith, Iowa Manufacturing Co. and others) consist of a stationary jaw and a swing jaw mounted directly on the eccentric shaft so that it receives downward and forward motions. The lower end of the swing jaw is held in position against the toggle by a tension rod.

The disadvantage of the known crushers lies in the complex swing of the jaw that brakes the crushed material in the crusher outlet. The lower end of the swing jaw held by the toggle arrangement moves along an arc upward and downward with an alternative velocity and obstructs discharging the material, and thus decreases the throughput. The objective of the present-invention is to increase the throughput of the crusher by eliminating the above disadvantage.

SUMMARY OF THE INVENTION

The above objective is achieved by mounting an additional synchronized eccentric drive of the lower end of the swing jaw. At this, the stationary jaw can be also made a swing one and provided with a toggle arrangement. Also, the stationary jaw of the present crusher can be provided with arc extensions wrapped around either upper, or lower interacting eccentric hinges of the swing jaw, or around the both jaws and forming respectively the primary and secondary crushing inlet and outlet in the converging passage between them and interacting with each other as a hammer and anvil.

At this the vertical component of the jaw movement and the falling down mass of the jaw are also utilized, whereas at present only the horizontal component is used. This greatly increases the working displacement and crushing area, the mechanical drive working more uniformly, with less horsepower, dynamic load, mass and strength requirements, balancing and the foundation. To facilitate the throughput still more, the swing jaw can be provided with transfer tooth elements working as a scraper conveyor. Almost complete counterbalancing is achieved by the mutual eccentric drive of both swing jaws. Therein one jaw is provided with ears hinged in the eccentrics of shafts on which the second jaw is mounted directly on the second eccentric. Both eccentrics are situated 180° out of the phase.

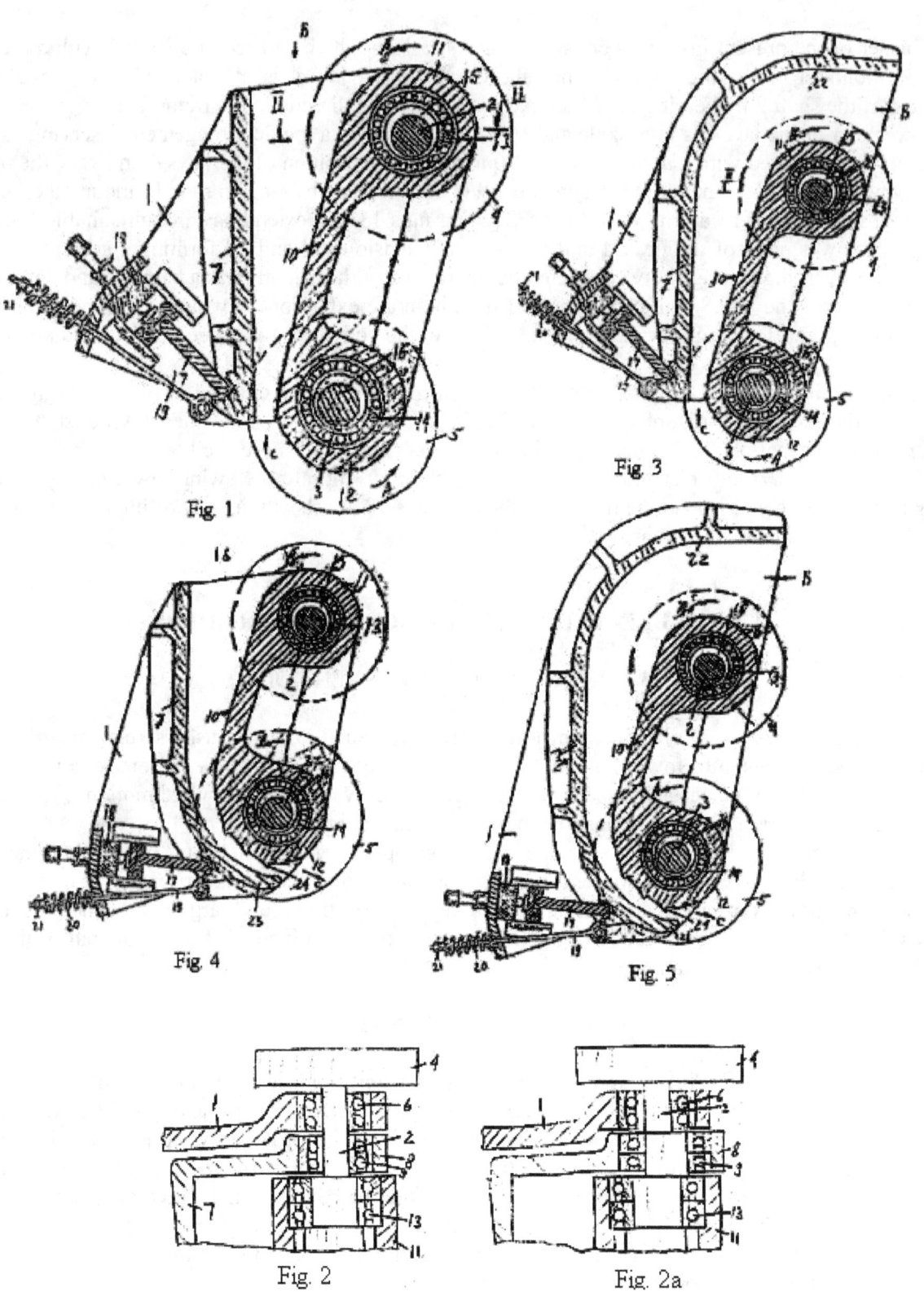

Fig. 1

Fig. 3

Fig. 4

Fig. 5

Fig. 2

Fig. 2a

IN THE DRAWINGS

Fig. 1, 3-5 are cross-sectional view of the first, second, third and fourth embodiments of the present invention respectively.

Fig. 2 and 2a are cross-sectional view taken along line II-II in Fig. 1, with one and both swing jaws respectively.

DESCRIPTION OF THE PREFERRED EMBODIMENTS

The present crusher consists of a frame 1 on which two synchronized eccentric shafts 2 and 3 with mechanical drives 4 and 5 are mounted in bearing supports 6 (the lower bearing support is not shown). A stationary jaw 7 is mounted with its hinge 8 on bearing 9 of the shaft 2 directly (Fig. 2). The swing jaw 10 is mounted with its hinges 11 and 12 on bearings 13 and 14 of the eccentrics 15 and 16. If the jaw 7 swings too (Fig. 2a), then it also hinged on eccentrics, both eccentrics being situated 180° out of the phase.

The lower part of the jaw 7 is releasable fixed in position by a toggle arrangement 17 serving as a safety mechanism and provided with a series of shims 18 adjusting the position of the jaw 7 to vary the discharge opening between the jaws. Tension rods 19 mounted on the frame 1 have springs 20 and threaded ends carrying nuts 21 and extended through suitable holes in the frame 1 hold the toggle arrangement 17 in place.

The jaw 7 can be provided with arc extensions 22 and 23 wrapped accordingly around either upper (Fig. 3) or lower (Fig. 4) eccentric hinges 11 and 12, or both (Fig. 5), and forming respectively the primary and secondary crushing inlet and outlet in the converging passage between them and interacting with each other as a hammer and anvil. Both synchronized shafts 2 and 3 rotate in the same direction shown by arrows A and with equal velocities. Therefore, the jaw 10 has plane-parallel and progressive motions.

The rock material fed into the crusher as indicated with arrow B is crushed between the jaws and in the converging passages, the jaw 10 pushing the material through. Still more throughput facilitating is achieved with transverse tooth elements 24 covering partially or entirely the swing jaw working surface and working as a scraper conveyor (in the drawing, the elements 24 are shown in way of illustration, but not in a limiting sense, on the hammer extension). The mutual eccentric drive of both swing jaws (Fig. 2a) counterbalances moving masses almost completely.

I.3. CAM CONE CRUSHER

BACKGROUND OF THE INVENTION

This invention relates to cone crushers including a conical bowl and a rotary cone. In such known crushers, an eccentric shaft or an unbalanced rotor drives the cone alternately toward and away from the bowl. Above crushers have the following disadvantages:

having only one local maximum convergence zone that travels through the passage between the cone and the bowl;

not using the rest of the bowl circumference at the same time;

unbalancing the cone.

The objective of the present invention is to eliminate above disadvantages and to elevate the efficiency of the crusher.

SUMMARY OF THE INVENTION

The above objective is achieved by executing the cross-section of the cone in the shape of a multi-profiled cam and in mounting the cone on an axial driving shaft. Such a construction increases capacity and the quantity of the crushing zones accordingly to the profiles quantity, completely balances the rotor and lessens the requirements to the strength of the bowl. However, the wear of the material engaging surfaces is greater.

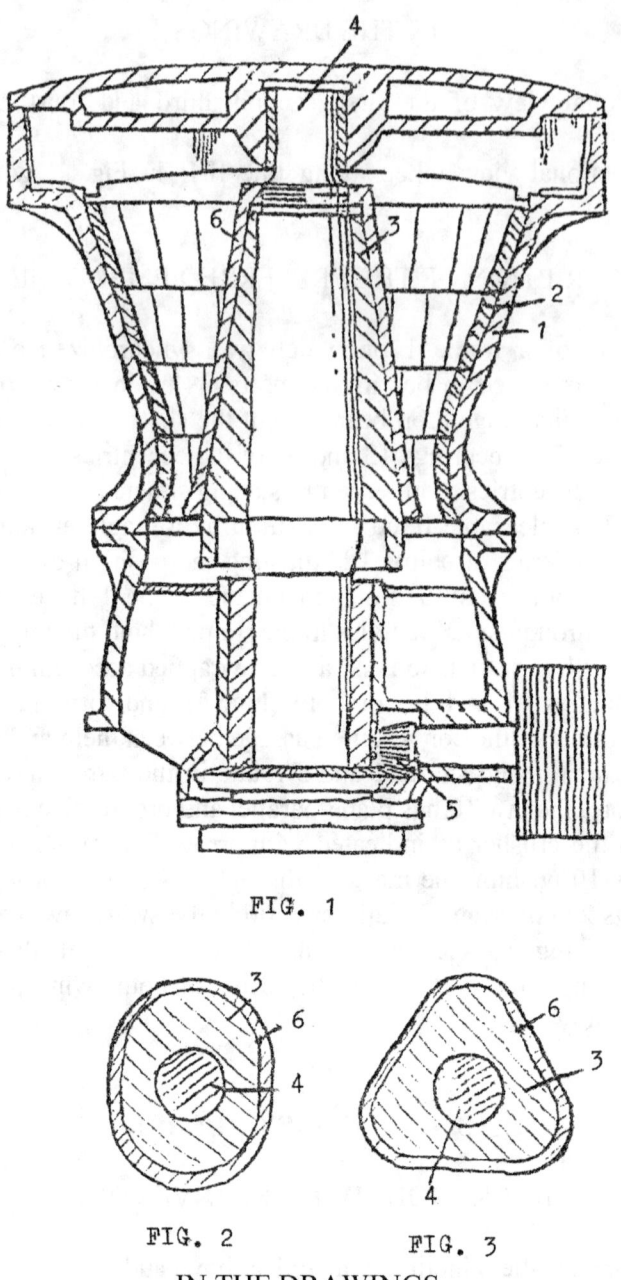

FIG. 1

FIG. 2   FIG. 3

IN THE DRAWINGS

Figure 1 is a cross-sectional view of the present crusher.

Figures 2 and 3 are respectively the cross-sectional view of the cone with double- and triple-profile cams.

DESCRIPTION OF THE PREFERRED EMBODIMENT

The present crusher consists of a bowl 1 having its crushing cavity lined with smooth or ribbed plates 2 of manganese steel. A rotary cone 3 is installed on a vertical shaft 4 driven a conical transmission 5. The shaft 4 is mounted through the bowl and cone mutual axis of the symmetry. The cone 3 is also provided with liners 6. The cross-section of the cone 3 is shaped as a double (Figure 2) or triple (Figure 3) profile cam.

When the cone rotates, the cam profile drives the local zones of maximum convergence through the passage between the jaws. Therefore, the crushed material falling down under the force of gravity is subjected to pressure in several zones instead of one (as in known gyratory crushers).

I.4. ARCUATED-JAW CRUSHER

BACKGROUND OF THE INVENTION

This invention relates to overhead eccentric jaw crushers. Such known crushers (e.g. manufactured by Kennedy Van Saun, Telsmith, Iowa Manufacturing Company and others) consist of a stationary jaw and a swing one mounted directly on the eccentric shaft so that it receives a downward-and-forward motion. The lower end of the swing jaw is held in position against the toggle by a tension rod. The objective of the present invention is to elevate the efficiency of such crushers and decrease their requirements to the construction strength, balancing and mechanical drive.

SUMMARY OF THE INVENTION

The above objective is achieved by providing the stationary jaw with an arc extension wrapped around the eccentric hinge of the swing jaw and forming a primary crushing inlet. This improvement allows utilizing the vertical component of the swing jaw working stroke, whereas the known crushers use only the horizontal component of the latter. Since the present crusher utilizes the gyratory movement of the eccentric, it can be compared with known gyratory and eccentric roll crushers of Rapidex Inc. and Joy Manufacturing Co.

The working cycle of the present crusher is almost doubled in comparison with conventional jaw crushers. Therefore, it works more uniformly with the less horsepower, dynamic load and mass of the crusher.

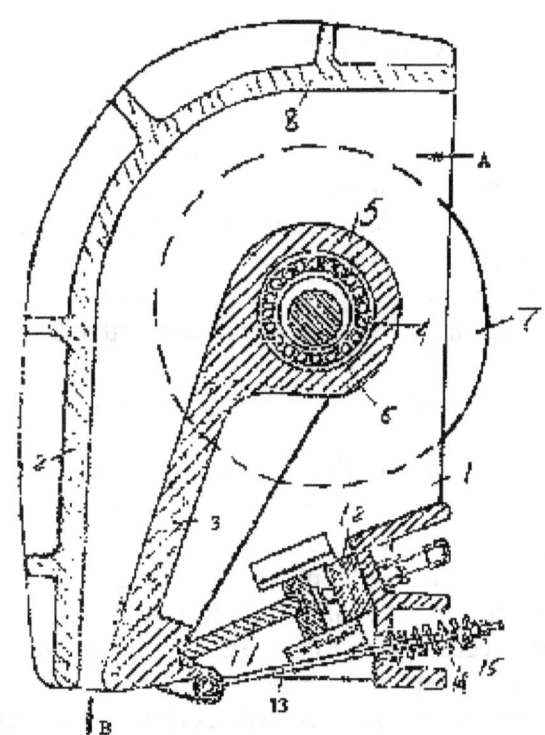

IN THE DRAWING

A lone figure is a cross-sectional view of the present crusher.

DESCRIPTION OF THE PREFERRED EMBODIMENT

The present crusher consists of a frame 1 to which a stationary jaw 2 is fixed, and a swing jaw 3 mounted in bearings 4 of its hinge 5 on an eccentric shaft 6. It is installed in bearing supports of the frame 1 (the supports are not shown) and provided with a mechanical drive 7. The upper part of the jaw 1 is has an arc extension 8 wrapped around the hinge 5 and forming a primary crushing chamber, or a converging passage between the extension 7 and the hinge 5. The continuation of the passage represents the chamber of a conventional jaw crusher itself.

The lower end of the jaw 3 is releasable and is fixed in position by a toggle arrangement 11 serving as a safety mechanism and provided with a series of shims 12 used to adjust the position of the jaw 3 to vary the discharge opening between the jaws. Tension rods 13 mounted on the frame 1 have springs 14 and threaded ends carrying nuts 15 and extended through a suitable hole in the frame 1 hold the toggle arrangement 11 in place.

A rock material is fed into the top of a converging passage between the extension 8 and the hinge 5 as indicated with arrow A and is moved downward into the space between the plain surfaces of the jaws. The material trapped in the converging passage is crushed by the vertical components of the movement of the hinge 5 to the extension 8. At this, a peristaltic pumping action enhances throughput of the material through the crusher. The material trapped between the jaws themselves is crushed by the horizontal component of the movement of the jaw 3 (as in conventional jaw crushers). The material is discharged through the opening between the jaws as indicated with arrow B.

Such a construction of the crusher improves its exploitation, decreases the dynamic forces and more uniformly loads its mechanical drive, increases its output and decreases its mass and requirements to the foundation.

I. 5. DOUBLE FED TWIN-JAW CRUSHER

BACKGROUND OF THE INVENTION

This invention relates to overhead eccentric twin jaw crushers. Such known crushers (e.g. manufactured by Iowa Manufacturing Company) consist of two synchronized swing jaws mounted directly on the eccentric shafts so that they receive downward and forward motions. Tension rods held the lower ends of the swing jaws in a position against the toggles. Twin jaw crushers are better balanced than single jaw ones, have 40% greater capacity and 2-4 times longer service life of their jaws.

The objective of the present invention is to elevate the efficiency of the twin jaw crushers still more and to decrease the requirements to the construction strength, balancing and mechanical drive.

SUMMARY OF THE INVENTION

Above objective are achieved by providing a stationary V-shaped jaw with two arcuated wings wrapped around the interacting eccentric hinges of the swing jaws and forming two oppositely situated primary crushing chambers. This improvement permits utilization of the vertical component of the swing-jaw working stroke, whereas at present, only the horizontal component of the latter is used.

Since the crusher of the present invention utilizes the eccentric's gyratory movement, it can be compared with known gyratory crushers and eccentric roll crushers of Rapidex Inc. and Joy Manufacturing Co. The working cycle is almost doubled in the present invention in comparison with conventional twin jaw crushers. Therefore, the present crusher works more uniformly with less horsepower, dynamic load and crusher's mass.

IN THE DRAWING

A lone Figure is a cross-sectional view of the present crusher.

DESCRIPTION OF THE PREFERRED EMBODIMENT

The crusher of the present invention consists of a frame 1 to which swing jaws 2 are mounted in bearings 3 of hinges 4 on eccentric shafts 5 installed in bearing supports (not shown) of the frame 1 and provided with synchronized mechanical drives 6. The upper part of the frame 1 has a stationary V-shaped jaw 7 with two arcuated wings wrapped around the interacting eccentric hinges 4 and forming two oppositely situated primary crushing chambers, or converging passages between the arcs and the hinges 4.

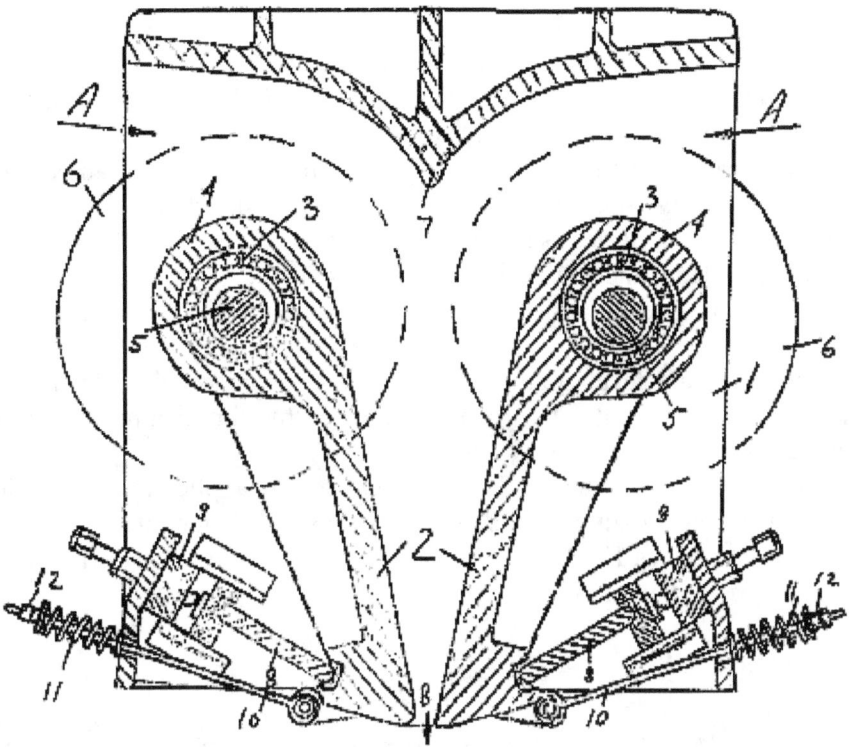

The lower ends of the jaws 2 are releasable fixed in position by toggle arrangements 8 serving as safety mechanisms and provided with a series of shims 9 adjusting the position of the jaws 2 to vary the discharge opening between the jaws. The toggle arrangements 8 are held in place by tension rods 10 mounted on the frame 1 and have springs 11 and threaded ends carrying nuts 12 and extending through suitable holes in the frame 1.

Rock materials enter into the top of the converging passages between the wings or the jaw 7 and the hinges 4, as indicated with arrows A, and moves downward between the jaws 2. The material trapped in the converging passages is crushed due to the vertical component of the hinges movement to the jaw 7. At this, a peristaltic pumping action enhances the material throughput through the crusher. The material trapped between the jaws 2 is crushed by the horizontal component of their movement (as in conventional twin jaw crushers). The material is discharged through the opening between the jaws 2 as indicated with arrow B.

The present crusher construction improves the conditions of its exploitation, decreases the dynamic force, loads its mechanical drive more uniformly, increases the output and decreases the crusher's mass and foundation requirements.

I.6. COIL CONE CRUSHER

BACKGROUND OF THE INVENTION

This invention relates to crushers including a conical bowl and a cone driven away or toward each other. In such known crushers, the bowl is made of relatively rigid continuous construction, the cone being driven by an eccentric shaft, unbalanced rotor, etc. This is accompanied with heavy masses of the construction, mechanical drive and foundation. The objective of the present invention is to eliminate the above disadvantages.

SUMMARY OF THE INVENTION

Above objective is achieved by the conical bowl shaped as a helical coil of an elastic spring material. One base end of the coil is fixed to the crusher frame. Other end is provided with a drive of torsional or linear oscillations.

Thus, the bowl of the present invention represents a conical spring. Therefore, if the coil of the spring has a uniform cross-section of its turns, they change their diameter under said oscillations gradually according to the different stiffness of the turns of the different diameters. That is why such a crusher produces only one local zone of maximum convergence traveling through the passage between the spring and cone on the helicoidal trajectory. This greatly reduces the required gearing ratio of the mechanical drive and mechanical stresses of the construction. Besides, since metal susceptibility to tensile stresses is much better than that to a transverse load and compressing, the spring is relatively stronger.

Above effects decrease decidedly the mass and dimensions of the crusher and simplify its construction. Also, such a spring design provides a peristaltic pumping action which increases the throughput of the crusher.

If it is necessary to increase said local zone of maximum convergence, the spring can be made of more uniform stiffness and constant spring rate. This can be achieved by diminishing the cross-sections of spring turns from one end to the other.

This invention is made in line with the author's earlier spring applications published in SPRING IN PECULIAR CASES (In Russian. Pruzhina v Osobykh Sluchayakh. Izobretatel' i Ratsionalizator (Inventor and Rationalizer/Innovator). Moscow, 1968, No. 3 (please see Addendum IV.1. Spring in Peculiar Cases).

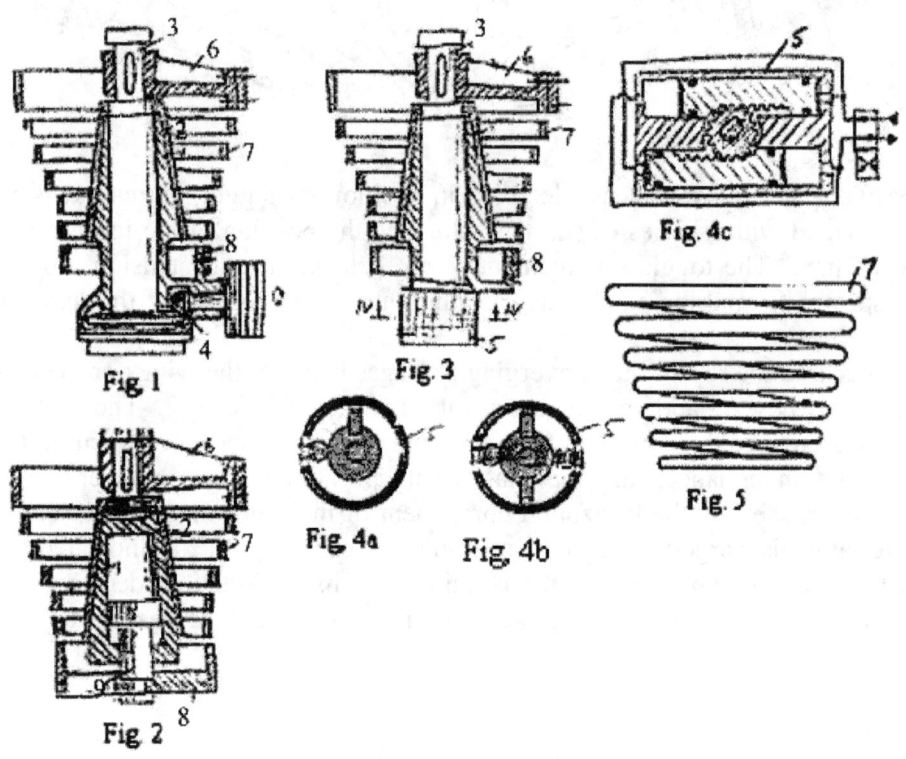

Fig. 1

Fig. 3

Fig. 4c

Fig. 4a

Fig. 4b

Fig. 5

Fig. 2

IN THE DRAWINGS

Fig. 1, 2 and 3 are the cross-sectional views of the present crusher appropriately with a mechanical drive of torsional oscillations, a hydraulic drive of linear oscillations and a hydraulic drive of torsional oscillations

Fig. 4a, 4b and 4c are cross-sectional views taken along line IV-IV in the Fig. 3 appropriately with a single-vane actuator, a double-vane actuator and a double rack-and-pinion actuator.

Fig. 5 is an appearance view of the spring bowl of the relatively uniform stiffness and spring rate.

DESCRIPTION OF THE PREFERRED EMBODIMENTS

The present crusher is based on a hollow cone 1 provided with a liner 2 and incorporating a mechanical (Fig, 1) or hydraulic (Fig. 2 - 4) drive of torsional (Fig. 1, 3, 4) or linear (Fig. 2) oscillations. In the cavity of the cone 10, a shaft 3 is installed (Fig. 1 and 3). The shaft 3 is driven by a conical gear 4 (Fig. 1) or hydraulic actuator 5 with a single (Fig. 4a) or double (Fig. 4b) vane, or a rack-and-pinion actuator (Fig. 4c). The other end of the shaft 3 is coupled with an arm 6 fixed to a base end of a conical spring 7 representing a bowl. In the crusher of Fig. 2 (which does not have the shaft 3), the arm 6 is coupled to the cone. The second base end of the spring 7 is fixed to an arm 8 joined to the cone (Fig. 1 and 3) or fixed to a plunger 9 of the hydraulic cylinder of the cone cavity (Fig. 2). All shown hydraulic actuators are of conventional design and do not require explanations, the main functional difference lying in the degree of rotation.

The processed material falls down under gravitation into the openings between the spring 7 and cone 1, and between the spring turns. The diameter of the spring turns is decreased when the spring is twisted by the torsional drive, and increased when the spring is unwound. The linear oscillations of the hydraulic cylinder (Fig. 2) initiate also the torsional component of the oscillations. Therefore, the material is pressed by the spring to the cone in only one local zone traveling on the spring helix.

At diminishing the cross-sections of the spring turns from one base to the other (Fig. 5), the spring has more uniform stiffness and spring rate. Such spring increases the said zone, but requires more torque and gearing ratio. At the same time, conical springs of conventional designs exhibit a non-linear force versus deflection characteristics due to the different spring rates and stress limits from coil to coil.

I.7. TOOTHED GYRATORY CRUSHER

BACKGROUND OF THE INVENTION

This invention relates to gyratory crushers comprising a bowl and a gyratory cone driven alternatively toward and away from each other.

The disadvantage of such known crushers lies in subjecting the shaft and bearings of the driven cone to the crushing pressure directly, without any kinematic linkages. Known double-toggle crushers do have a kinematic linkage between a crushing member (a jaw) and a driving shaft. This linkage represents the double-toggle mechanism taking the crushing load and providing the second stage of mechanical amplification. Therefore, such crushers are suitable for the hardest materials.

The objective of the present invention is to eliminate the above disadvantage.

SUMMARY OF THE INVENTION

The above objective is achieved by means of that the cross-section of' said bowl and cone constitutes internal tooth engagement representing a plurality of double-acting jaw crushers on the circumference. Thus, the load is mostly closed between the teeth, the secondary mechanical amplifier having a wedge action.

IN THE DRAWINGS

FIG. 1 is a longitudinal sectional view of the present crusher;
FIG. 2 is a schematic cross-sectional view of the present crusher with a three-toothed cone;
FIG. 3 is the same as above with a four-toothed cone;
FIG. 4 is the same an above with a six-toothed cone;
FIG. 5 is the same as above with an eight-toothed cone.

DESCRIPTION OF THE PREFERRED EMBODIMENTS

The present crasher consists of a bowl 1. Its cavity is lined with plates 2 of a wearable material, e.g.

manganese steel. The bowl 1 has a cross-member 3 that has a spherical support 4. On the latter, a vertical shaft 5 with a fixed cone 6 is mounted. Said cone can be provided with liners 7. The shaft 5 is eccentrically suspended within an eccentric 8 in such a manner that the rotation of the latter by a conical transmission 9 causes the shaft 5 to gyrate relatively to the bowl 1. The cross-section of the crusher constitutes an internal tooth engagement and represents a plurality of double-acting jaw crushers on the circumference.

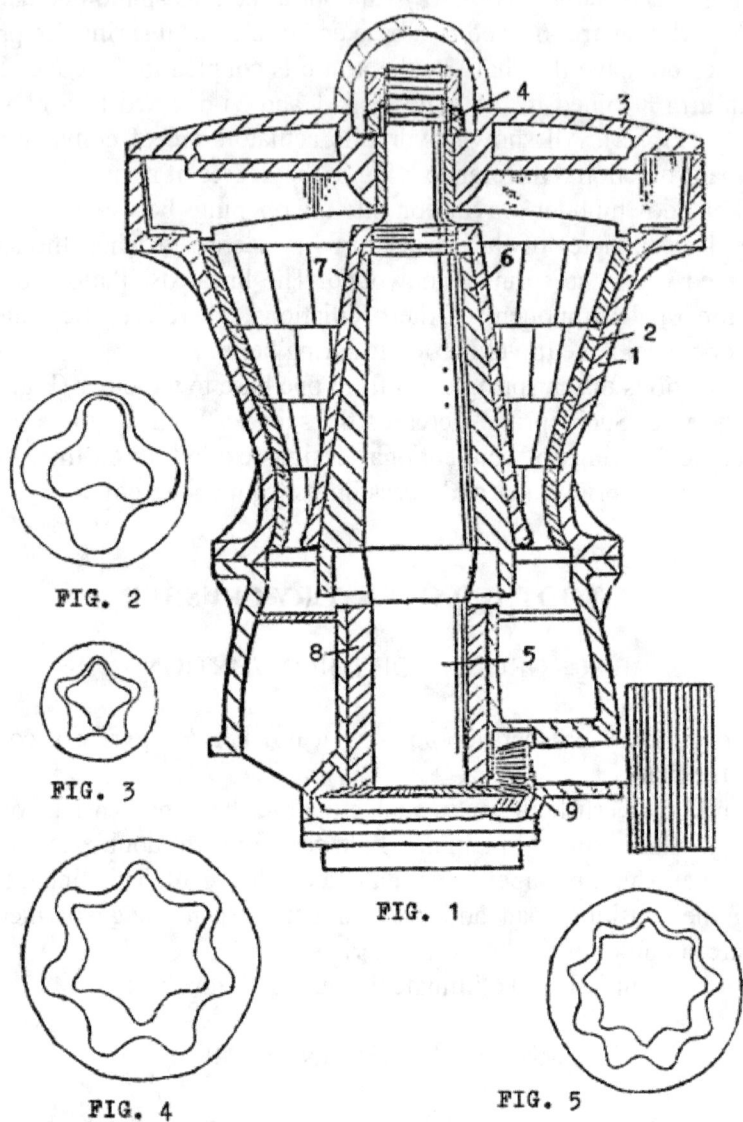

FIG. 2

FIG. 3

FIG. 1

FIG. 4

FIG. 5

FIG.3-5 show examples with 3, 4, 6 and 7 teeth cones accordingly. Therein the cone stands for the inner gear and the bowl – for the outer one. To lessen the relative sliding speed of the gears, the bowl can have only one tooth more than the cone.

The rotating eccentric 8 swings the cone 6 about its vertical axis. The cone approaching the bowl crushes the rock material to the size small enough for the falling down by gravitation through the discharge opening when the cone moves away. The teeth work like wedges filling the clearance space in the improvised gearing and provide a wedge amplifier.

Obviously, many modifications and adaptations can be made without departing from the spirit and scope of the present invention.

I.8. HORIZONTALLY FED SINGLE-SWING-JAW CRUSHER

BACKGROUND OF THE INVENTION

This invention relates to horizontally fed jaw crushers comprising an outer jaw with a concave curved material engaging surface, an inner jaw with an outer material engaging curved surface, and an eccentric drive of one jaw. In such known crusher (U.S. Patent No. 4,165,042), the inner jaw has its eccentric drive situated under the outer jaw and the radial fed. The objective or the present invention is to achieve a better peristaltic pumping action in the crusher for increasing its throughput.

SUMMARY OF THE INVENTION

Above objective is achieved by dividing the outer jaw into several separate sections situated under the inner jaw and provided with ears interacting with eccentrics spaced on the shaft in such a manner that a progressive transverse undulating movement about the inner jaw is created progressively from the crusher's inlet to its outlet.

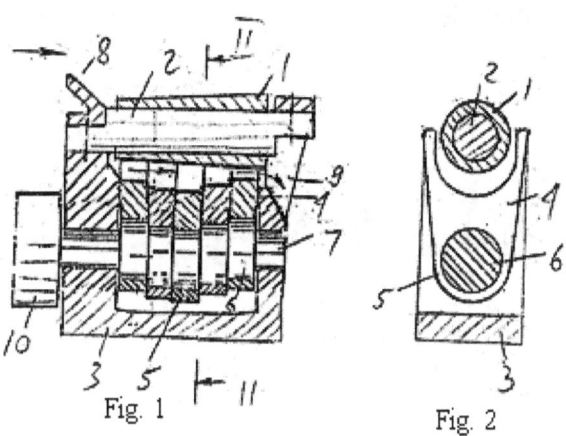

Fig. 1   Fig. 2

IN THE DRAWINGS

Fig. 1 is a cross-sectional view of the present crusher.
Fig. 2 is a cross-sectional view taken along line II-II on Figure 1.

DESCRIPTION OF THE PREFERRED EMBODIMENT

An inner jaw of the present crusher consists of a hollow cone 1 installed on a rod 2 supported by a frame 3. The discharge width opening is adjusted by moving the rod 2 with the cone 1 along the frame 3. A safety mechanism for releasing unbreakable materials can be executed in a manner similar to the spring mechanisms of known roll crushers or jaw crushers (not shown).

The outer jaw is divided into sections 4 situated under the inner jaw and provided with ears 5 interacting with eccentrics 6 spaced on a shaft 7 in such a manner that a progressive transverse undulating movement about the inner jaw is created progressively from the crusher's inlet 8 to its outlet 9. The sections 4 have a concave curved material engaging surface. A mechanical drive 10 drives the shaft 7.

Rock material is fed into the inlet 8, crushed when the sections 4 approach the cone 1 and propagated to the outlet 9 by the peristaltic pumping action of the rotating shaft 7. Still better pumping action is achieved if the shaft 7 has two cranks (in Fig. 1 one crank is shown in way of illustration, but not in a limiting sense). In this case, a relatively closed chamber is created between the cranks describing 360°. However, such an embodiment is more complicated.

The illustrated embodiment can be varied in different ways within the scope of the invention. Thus, the cone and rod can be replaced with a conventional jaw, with another material engaging surface, etc.

I.9. HORIZONTALLY FED DUAL-SWING-JAW CRUSHER

BACKGROUND OF THE INVENTION

This invention relates to horizontally fed jaw crushers comprising an outer jaw with a concave curved surface and an eccentric drive. In such known crusher (U S Patent No. 4,165,042), the inner jaw is situated under the outer jaw and fed radially direction. The disadvantages of the known crusher lies in its cyclical work (i.e. the swing jaw makes one power stroke and one idling stroke during one revolution of the eccentric shaft) with higher requirements to the strength of the construction, balancing and mechanical drive; a relatively small area for interacting jaws and wear of the outer jaw.

The objective of the present invention is to eliminate the above disadvantages and achieve better peristaltic pumping action in the crusher for assisting its throughput.

SUMMARY OF THE INVENTION

The above objective is achieved by dividing the jaws into separate sections, the outer jaw being situated under the inner jaw and provided with ears interacting with eccentrics of the second shaft synchronized with the eccentric drive of the inner jaw. The eccentrics are spaced on said shafts in such a manner that a progressive transverse undulating movement is created progressively from the inlet to the outlet of the crusher.

IN THE DRAWINGS

Fig. 1 is a cross-sectional view of the crusher of the present invention.
Fig. 2 is a cross-sectional view taken along line II-II in Fig. 1.

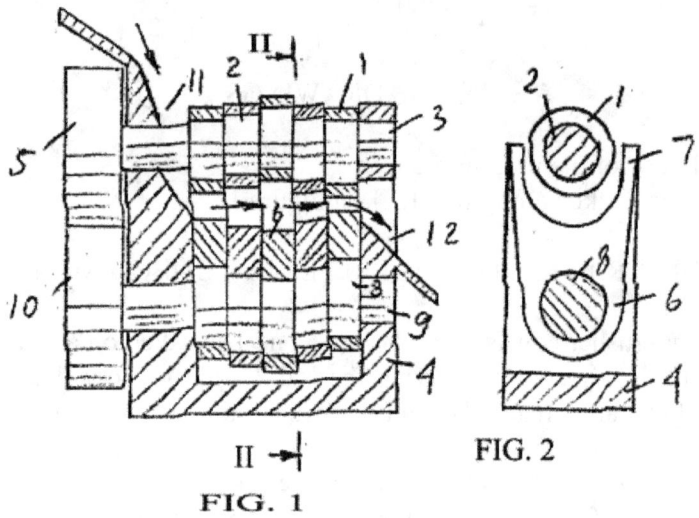

FIG. 1

FIG. 2

DESCRIPTION OF THE PREFERRED EMBODIMENT

The crusher of the present invention comprises an inner jaw representing several rolls 1. They are mounted on eccentrics 2 of a shaft 3 installed on a frame 4 and provided with a mechanical drive 5. An outer jaw has the adequate quantity of sections 6 situated under the shaft 3 and equipped with ears 7. The latter interact with eccentrics 8 of the second synchronized shaft 9 driven in the opposite direction by a mechanical drive 10. The eccentrics 2 and 7 are spaced on the shafts 3 and 9 in such a manner that a progressive transverse undulating movement is created progressively from an inlet 11 to an outlet 12 of the crusher. The sections 6 have a concave curved material engaging surface.

A safety mechanism for releasing a non-breakable material and adjusting the width of the discharge opening can be executed in a manner similar to the spring mechanisms of known roll or jaw crushers (not shown).

Rock material is fed into the inlet II, crushed when the rolls 1 and sections 6 approach each other and is propagated to the outlet 12 under peristaltic pumping action.

It is obvious that better pumping action is achieved if the shafts 3 and 9 have two cranks (in Fig. 1 one crank is shown in way of illustration, but not in a limiting sense). In this case, a relatively closed chamber is created between the cranks describing 360°. However, such an embodiment is more complicated and is not necessary for loose friable materials. In the present embodiment, an inclined crushing chamber sloping downwardly from the inlet to the discharge provides the effect of sealing, the material not moving upwardly.

I.10. ROLL JAW CRUSHER

BACKGROUND OF THE INVENTION

This invention relates to crushers provided with eccentric rolls acting against a stationary anvil (jaw). Such a known crusher (C.R. Peterson and Allan T. Fisk. Portable Low Profile Crusher for Underground Mining Applications, Rapidex Inc. Report for Bureau of Mines, Department of the Interior, Washington, D.C. 20241) consists of a rotating horizontal eccentric placed inside a bearing-supported free turning roller (named a rotary jaw) and a fixed wrap outer jaw (an anvil, if to apply the known roll crushers' term). Crushed rocks fed into the horizontal inlet are discharged by the gravity at the lower side.

Denver Equipment Division of Joy Manufacturing Co. manufactures a similar crusher. Although Rapidex relates their crusher to the jaw ones, the principal distinguishing characteristics lying in the so-called rotary jaw and in the arc shape of the stationary jaw. Such a crusher can be also related to roll crushers, the principal distinguishing characteristics lying in the eccentric disposition of the roller on the rotating shaft, and to gyratory crushers, the principal distinguishing characteristics lying in the cylindrical shape of working members and in utilizing only 1/3 of the circular surface of the gyratory crushers' bowls. However, from the point of view of the mechanics of interacting crushing members, the gyratory crushers are the closest ones.

The disadvantage of the Rapidex crusher in comparison with known gyratory ones lies in its cyclical work, i.e. the roller makes one working stroke and one idling stroke during one revolution of its eccentric shaft. This commands high requirements to the construction strength, balancing and to the mechanical drive.

The objective of the present invention is to eliminate those disadvantages.

SUMMARY OF THE INVENTION

The above objective is achieved by dividing the roll into several separate sections situated on a multi-eccentric shaft with its eccentrics spaced equally apart each other on the shaft circumference. Such a crusher achieves continuous crushing of the material since the separate roll sections work in turn, the more sections, the more uniformly it works. Its different rollers achieve squeezing and releasing the material simultaneously. Such uniform work requires less horsepower, develops less dynamic load and decreases the crusher's mass.

IN THE DRAWINGS

Fig. 1 is a cross-sectional view of the crusher of the present invention.
Fig. 2 is a sectional view taken along line II - II in Fig. 1.

DESCRIPTION OF THE PREFERRED EMBODIMENT

The crusher of the present invention consists of a frame 1 to which a stationary jaw 2 (an anvil) is fixed and a multi-eccentric shaft 3 supported in bearings 4 in the frame 1 and provided with a mechanical drive 5. A liner 6 of a wear-resistant material, for example, manganese steel, can protect the working surface of the jaw 2 in a conventional manner. The shaft 3 is provided with two or more rolls 7 (in Fig. 2, three rolls are shown only in way of illustration, but not in a limiting sense) situated on eccentric collars 8 fixed to the shaft 3 with a key or a similar joint 9.

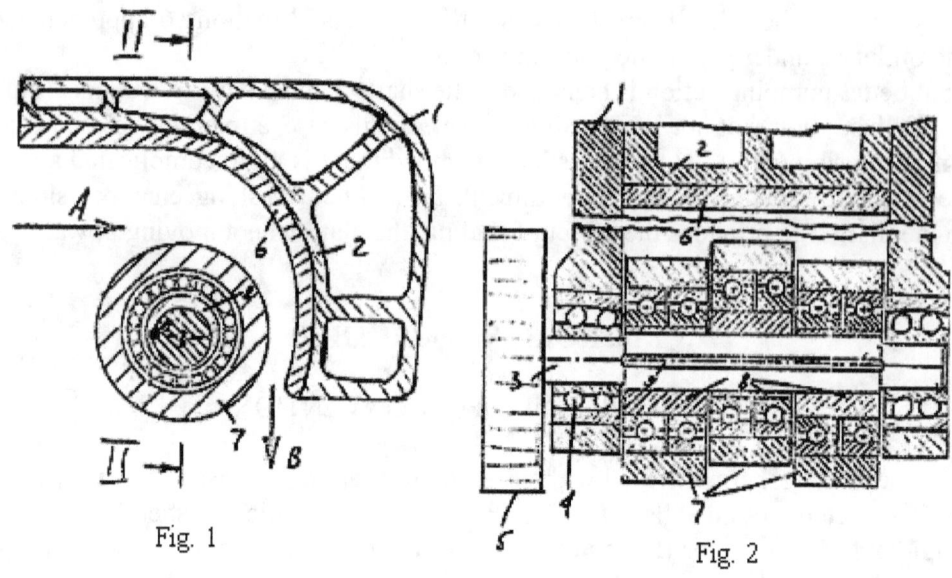

Fig. 1   Fig. 2

Rock materials are fed into the top of a converging passage between the rolls 7 and the jaw 2, as indicated with arrow A, and are moved downward. Further, the materials are fed by gravity and fall forward as the clearance between the rolls 7 and jaw 2 increases. The material trapped in the converging passage is crushed when the rolls 7 move to the jaw 2. The rolls 7 provide also a peristaltic pumping action that enhances the materials throughput as indicated with arrow B.

Such a construction improves the conditions of its exploitation, decreases the dynamic load and uniformly loads its mechanical drive, increases its output and decreases its mass and requirements to the foundation. Besides, the efficiency of the breaking down rocks is elevated since a transverse bending load on the rocks between the different rolls is probable.

I.11. MULTI-ROLL CRUSHER

BACKGROUND OF THE INVENTION

This invention relates to roll crushers. Two and three roll crushers with the opposite-direction rotating rolls pressing and rubbing the material are known (e.g. those of Iowa Manufacturing Co.). Such known crushers have lost favor to gyratory and jaw crushers for coarse crushing.

Crushers provided with the eccentric roll acting against a stationary anvil (jaw) are also known (e.g. those of Denver Equipment Division of Joy Manufacturing Company and U.S. patent No.4, 165,042).

Although the crusher in the above patent is named a rotary jaw crusher, it can be also related to roll crushers. The principal distinguishing characteristic is the eccentric disposition of the roller on the rotating shaft. Besides, that crusher relates to gyratory crushers too. The principal distinguishing characteristics is the cylindrical shape of working members and in utilizing only 1/3 of the circular surface of the bowl of gyratory crushers. However, from the point of view of mechanics of interacting crushing members, the gyratory crushers are the closest one.

The objective of the present invention is to provide greater capacity, better balancing, decreasing mass and longer service life of the working members.

SUMMARY OF THE INVENTION

In the first embodiment of the present invention, two interacting synchronized eccentric rolls achieve the above objective with their eccentrics mounted in the same phase and rotating in the opposite directions. In the second embodiment, said eccentrics are situated 180° out of the phase and rotate in the same direction.

The crushing is enhanced with the third eccentric roll situated above (the third embodiment) or underneath (the fourth embodiment) said two interacting eccentric rolls. In addition, still more enhancing is possible with four (the fifth embodiment) or more (not shown) interacting eccentric rolls. Whereas the two-roll embodiments have one crushing stage, the three, four or more roll embodiments have accordingly the second, third or more crushing stages.

Also, eccentric roll crushers can be provided with a stationary arc v-shaped jaw with its arc wings wrapped around the interacting eccentric rolls (the sixth embodiment) and situated above, beneath or in both the places.

Many other variants are possible in the scope of the present invention, e.g. combinations of eccentric rolls and jaws above, or beneath the two rolls, etc.

IN THE DRAWINGS

Fig. 1 is a cross-sectional view of the present two-roll eccentric crusher with synchronized eccentrics.
Fig. 2 is the same as above, the eccentrics situating 180° out of the phase.
Fig. 3 is a cross-sectional view of a three-roll eccentric crusher with the below third roll.
Fig. 4 is the same as above with the top third roll.
Fig. 5 is a cross-sectional view of a four-roll eccentric crusher.
Fig. 6 is the same as in Fig, 1 with a stationary jaw at the top of the rolls.
Fig. 7 is a cross-sectional view taken along line - VII - VII in Fig. 6.
Fig. 8 is the same as in Fig. 3 with a stationary jaw beneath the rolls.
Fig. 9 is the same as in Fig. 4 with a stationary jaw at the top.
Fig. 10 is the same as in Fig. 1 with two stationary jaws.

DESCRIPTION OF THE PREFERRED EMBODIMENTS

The present crusher consists of a frame 1 with eccentric shafts 2 and 3 supported in bearings 4 and having a mechanical drive 5. The crushers in Fig. I, 2, 6 and 10 have two eccentric shafts, the crushers in Fig. 3, 4, 8 and 9 have the third shaft 6 beneath the above two (Fig. 3 and 9) or an upper shaft 7 at the top (Fig. 4 and 8), the crusher in Figure 5 has four eccentric shafts.

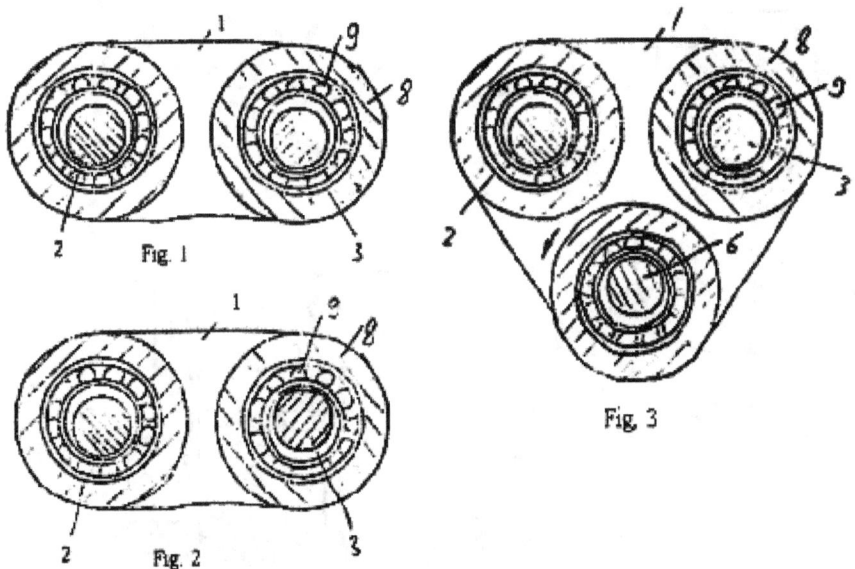

Fig. 1

Fig. 2

Fig. 3

On the eccentric of each shaft, a roll 8 is mounted with the possibility of free rotation (a bearing 9 for the roll 8 is not compulsory and is shown in way of illustration, but not in a limiting sense). The eccentrics of shafts 2 and 3 can be synchronized and rotating in the opposite directions (Fig. 1, 3 - 6, 8 - 10), or situated 180° out of the phase and rotating in the same direction (Fig. 2).

In the first case, the frame 1 is subjected to vertical linear oscillations, in the second case, to torsion oscillations with two times less dynamic load on its foundation. The throughput in the first case is more, however, the material is processed more thoroughly in the second case because of some braking action of the second roll.

Rocks fed into the clearance between rolls are crushed like in conventional jaw crushers, when rolls are moving to each other, and fall down when they are going back, crushed again on the next power stroke until they finally drop out as chips.

Three eccentric roll crushers provide twice the reduction ratio of a dual-eccentric roll crusher. The third roll is synchronized with the main rolls and can rotate in the direction of one or another main shaft depending on the desirable effect and feed quantity (with upper additional shaft) or outlet quantity (with the lower additional shaft).

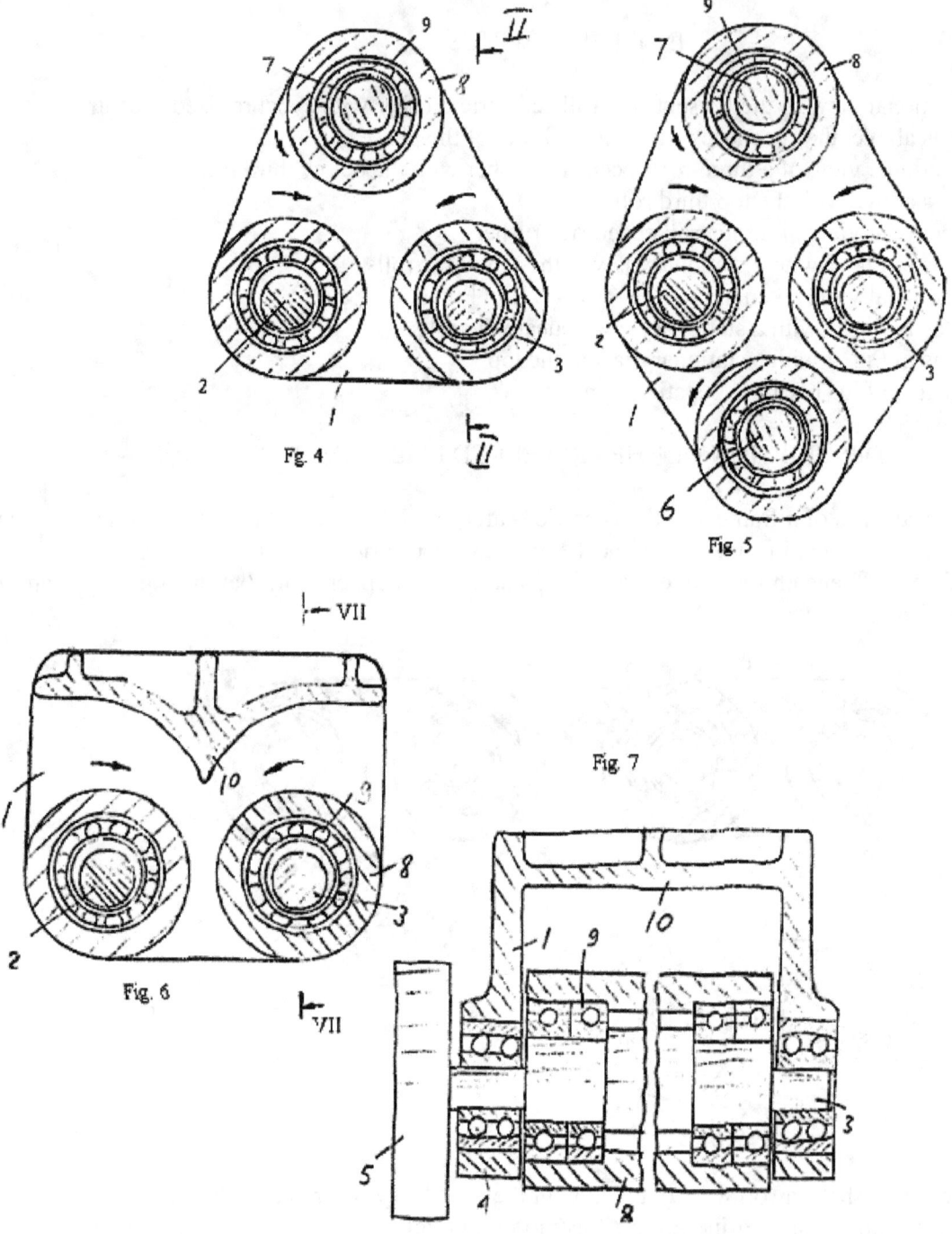

Fig. 4

Fig. 5

Fig. 7

Fig. 6

A single feed three-roll crusher has greater capacity and a smaller percentage of oversize as its second stage does not have to re-crush material reduced to finished size by the first stage. Dual feed for the crushers with the upper additional roll provides higher capacity. Both the feeds can be set differently (e.g. coarse setting and finer feed) or in the same manner. With the lower additional roll, dual outlet is possible with similar alternatives. The four-eccentric-roll crusher (Fig. 5) has three crushing stages and therefore provides still more capacity, dual feed and outlet being possible.

With stationary arc V-shaped jaws 10 and 11 are installed instead of either the upper (Fig. 6 and 9) or lower (Fig. 8) roll, or both of them (Fig. 10) and having their arc wings wrapped around the interacting rolls, the crushing chamber is formed in the converging passages between the arc and rolls. Such a crushing chamber works in the same manner as the known rotary jaw crushers. The jaw increases the working area, but might brake the material.

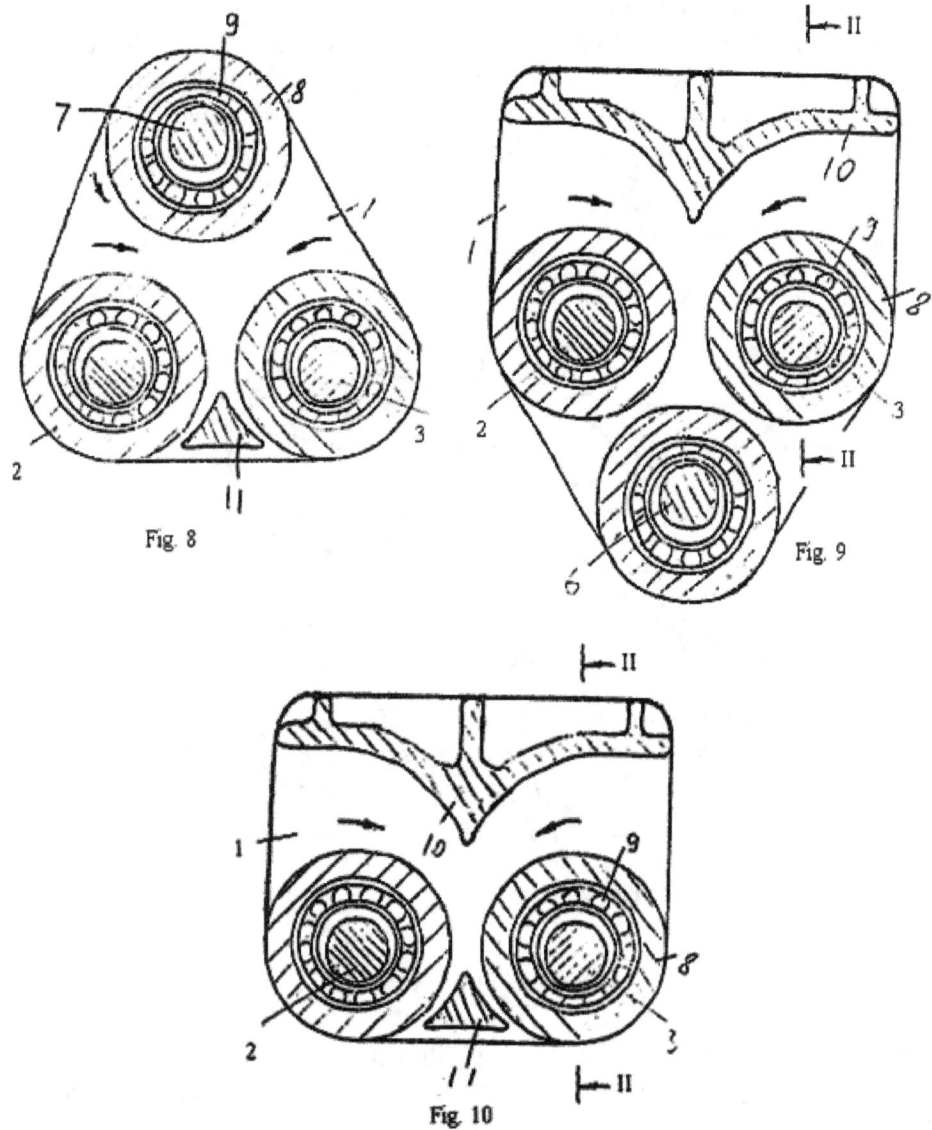

Fig. 8

Fig. 9

Fig. 10

Although nine variants are shown in the drawings, still more modifications can be made in arranging the jaws and rolls within the scope of the present invention.

I.12. CAM JAW CRUSHER

BACKGROUND OF THE INVENTION

This invention relates to rotary jaw crushers. Such known crushers. (e.g. US patent Nos. 2,464,732;

2,675,970; 1,946,763; 3,229,922 and 4,165,042) consist of a stationary outer concave jaw and a rotary inner jaw with an eccentric drive. Said jaws have curved material engaging surfaces: concave for the stationary jaw and convex for the rotary one.

Disadvantages of the known crushers lie in their cyclical work, i.e. the rotary jaw makes one working stroke and one idling stroke of the same duration during each revolution of its eccentric shaft, the rotor being unbalanced. The objective of the present invention is to eliminate the above disadvantages.

SUMMARY OF THE INVENTION

The above objective is achieved by shaping the rotary jaw as a cam with one or more profiles. For different durations of strokes, said profiles can be made unsymmetrical. Mono-profile cams give only different duration of the strokes, whereas the double or triple ones increase the quantity of the cycles accordingly and completely balance the rotor. Therefore, such a construction lessens the requirements to the frame strength, balancing and mechanical drive, and eliminates heavy flywheels of the known crushers. To decrease friction and wear of the cam, the latter can be provided with a floating ring liner or a roll-holder.

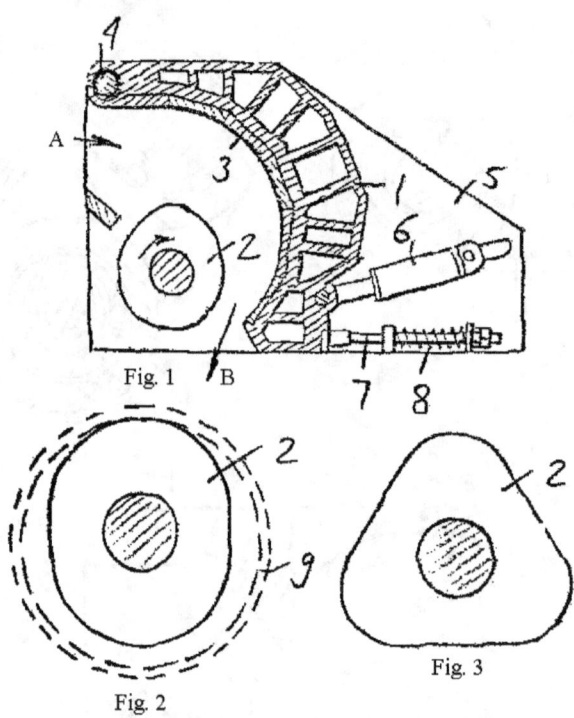

Fig. 1

Fig. 2

Fig. 3

IN THE DRAWINGS

Fig. 1 is a cross-sectional view of the present crusher with a mono-profile cam.

Fig. 2 and 3 are cross-sectional views of the rotor of the present invention appropriately with double- and triple-profile cams.

DESCRIPTION OF THE PREFERRED EMBODIMENT

The present crusher consists of a stationary outer jaw 1 (Fig. 1) and an inner smaller rotary cam jaw 2. Between the jaws 1 and 2, the material to be crushed passes. The jaw 1 has a concave curved material engaging surface 3 provided with a liner of a high-resistant material. The outer material engaging surface of the cam can be also provided with a liner (not shown). The jaw 1 is swung on a pivot joint 4 of a frame 5 and held in place by a hydraulic cylinder 6 and a rod 7 with a spring 8. Such a construction allows the release of non-breakable material and permits changing the opening of the crusher. These features are close to conventional designs and are self-explanatory.

The processed material is passed in the direction shown by arrow A and discharged in the direction shown by arrow B, the rotor rotating in the clockwise direction. The cam profile produces a local maximum convergence zone traveling through the passage between the jaws. Fast loading and slow-applied pressure can be achieved with an unsymmetrical profile of the cam. Multi-profile cams (Fig. 2 and 3) achieve a uniform force spread.

To prevent the excessive wear of the cam, a floating ring 9 (shown in Fig. 2 with dashed lines) or a roll holder (not shown) can be provided. However, such details decrease the throughput of the crusher.

I.13. MULTI-ECCENTRIC JAW CRUSHERS

BACKGROUND OF THE INVENTION

This invention relates to eccentric jaw crushers. Such known crushers (e.g. manufactured by Kennedy Van Saun, Telsmith, Iowa Manufacturing Co. and others) consist of a stationary jaw and a swing jaw mounted either directly on the eccentric shaft or pivoted at the top or the bottom.

In the first type, the swing jaw receives a downward-and-forward motion from the eccentric directly. In the second type, a knuckle action of the rising and falling second lever (pitman) carried by the eccentric shaft moves the jaw by two plates (toggles).

The disadvantage of the jaw crushers in comparison with known gyratory crushers, lies in their cyclical work, i.e. the jaw makes power and idle strokes during one revolution of the eccentric shaft. That increases the requirements to the construction strength, balancing and mechanical drive. Therefore, the objective of the present invention is to eliminate above disadvantage.

SUMMARY OF THE INVENTION

Above objective is achieved by dividing the swing jaw into separate sections situated directly on a multi-eccentric shaft. In another embodiment, the knuckle action of the pitman pivots said sections at either the top or the bottom. A multi-eccentric shaft carries the pitman. The eccentrics are spaced equally apart from each other on the shaft circumference.

The device crushes the material continuously since the separate jaw sections work in turn, the more sections the more uniformly the crusher works. The different jaw sections squeeze and release the material simultaneously. Such uniform work requires less horsepower, develops less dynamic load and decreases the mass of the crusher and foundation. Besides, the efficiency of the breaking rocks is elevated since transverse (bending) load on rocks between the different sections is probable.

IN THE DRAWINGS

Fig. 1 is a cross-sectional view of the overhead eccentric crusher.
Fig. 2 is a cross-sectional view of the twin overhead eccentric crusher.
Fig. 3 is a cross-sectional view of the pitman-type crusher.
Fig. 4 is a sectional view taken along line IV-IV of Fig. 1.

DESCRIPTION OF THE PREFERRED EMBODIMENTS

The present overhead eccentric jaw crusher (Fig, 1) contains a frame 1 to which a stationary jaw 2 is fixed, and a swing jaw 3. The latter is mounted on bearings 4 of its hinge 5 on a multi-eccentric shaft 6 that is installed in bearing supports 7 (Fig. 4) of the frame 1 and provided with a mechanical drive 8.

The swing jaw 3 is divided into several sections (in Fig. 4, three sections are shown only in way of illustration). They are placed on eccentric collars 9 fixed to the shaft 6 by a key or similar joint 10. The lower ends of the sections are releasably fixed in position by toggle arrangements 11 serving as safety mechanisms and provided with series of shims 12 adjusting the sections position to vary the discharge opening between the jaws. The toggle arrangements 11 are held in place by tension rods 12 mounted on the frame 1 and have springs 14 and threaded ends carrying nuts 15 and extended through suitable holes in the frame 1.

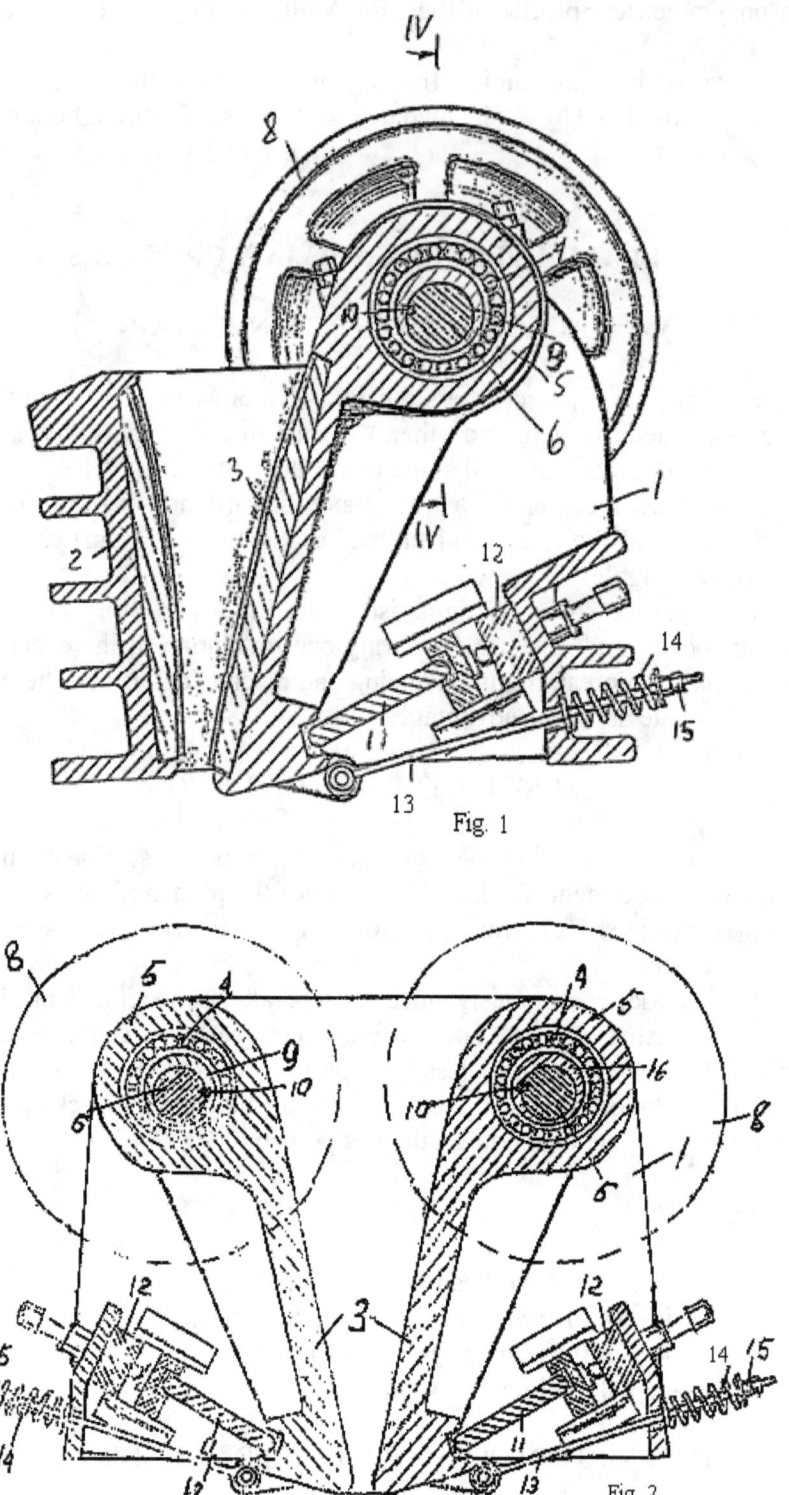

Fig. 1

Fig. 2

The twin-jaw crusher (Fig. 2) has the second swing jaw instead of the stationary one and is mirror aligned to and synchronized with the first jaw. In the pitman-type crusher (Fig. 3), the swing jaw 3 is swung on an axle 16, whereas the eccentric shaft 6 carries pitman arms 17 reciprocating in a vertical plane. The ends of toggles 11 are hinge-jointed to the lower part of the pitman arm 17.

Rocks are fed into the top of the crusher, broke by the sections of the jaw 3 when they moves to the opposite jaw and fall down when they are going back, crushed again on the next power stroke until they finally drop out as the chips through the discharge opening.

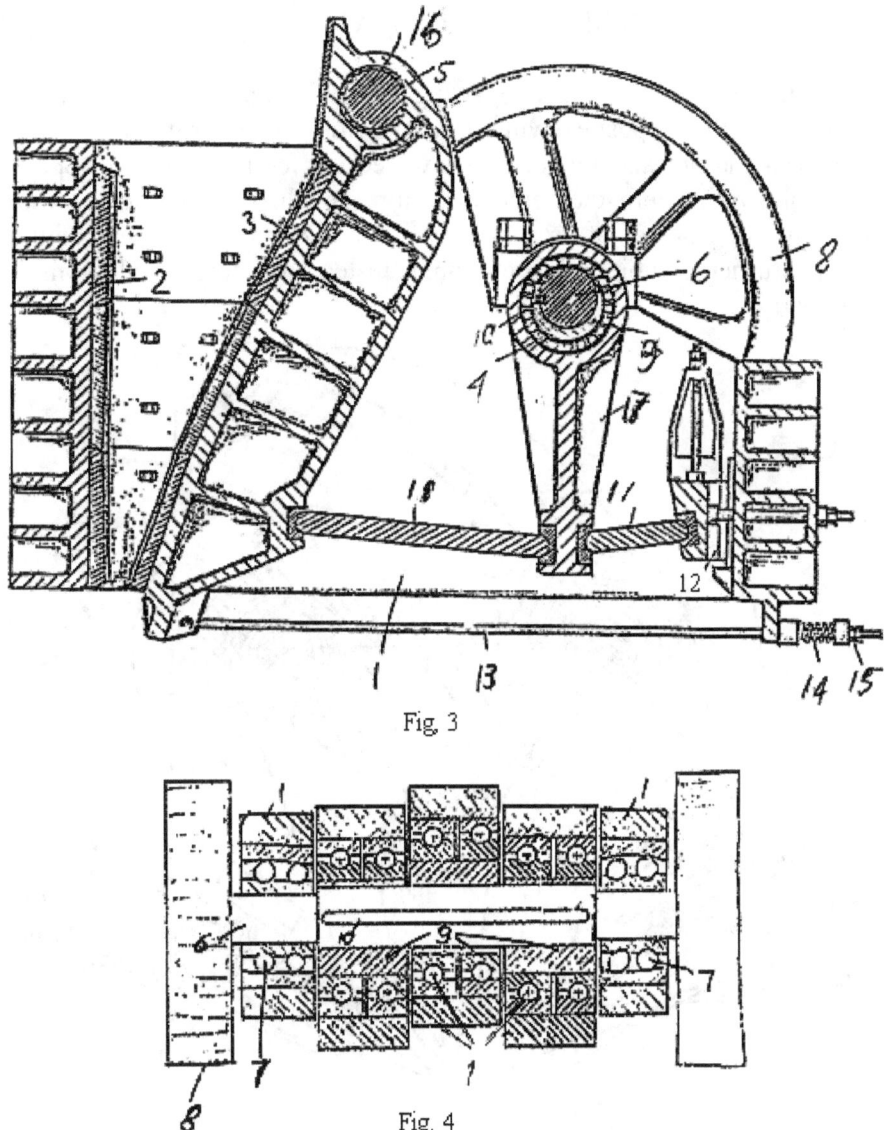

Fig. 3

Fig. 4

The resistance to the motion of the swing section increases sharply on the power stroke and diminishes on the return stroke. That is why several sections of the present crusher distribute the load on the mechanical drive, frame and foundation more uniformly and ensure the smooth running. Conventional balance wheels assist to the distribution, racing up, storing the surplus energy from the mechanical drive during idle stroke and releasing it for crushing on the power stroke. In addition, big rocks are subjected to transverse (bending) load between sections, which elevates the efficiency still more.

I.14. ECCENTRIC TWIN-JAWS CRUSHER

BACKGROUND OF THE INVENTION

This invention relates to overhead eccentric jaw crushers with both the swing jaws. Such known crushers (e.g. manufactured by Iowa Manufacturing Company) consist of two synchronized swing jaws mounted directly on the eccentric shafts providing a downward-and-forward motion. Tension rods hold the lower ends of the swing jaws in position against the toggles. The twin-jaw crushers are better balanced than single jaw crushers are, have 40% greater capacity and 2 - 4 times longer service life of their jaws (foreign data).

The objective of the present invention is to elevate the efficiency of the above crushers still more and to decrease the requirements to the construction strength, balancing and mechanical drive.

SUMMARY OF THE INVENTION

Above objective is achieved by providing one swing jaw with an arc extension wrapped around the interacting eccentric hinge of the opposite swing jaw and forming a primary crushing chamber in the converging passage between the arc and hinge. The jaws eccentrics are directed oppositely from the line connecting their centers and rotate into the same direction. By this, the vertical component of the jaws working stroke is also utilized, whereas at present only the horizontal component of the latter is used. In addition, the frame of the crusher is subjected to torsional (instead of linear) oscillations with two times less dynamic load on its foundation.

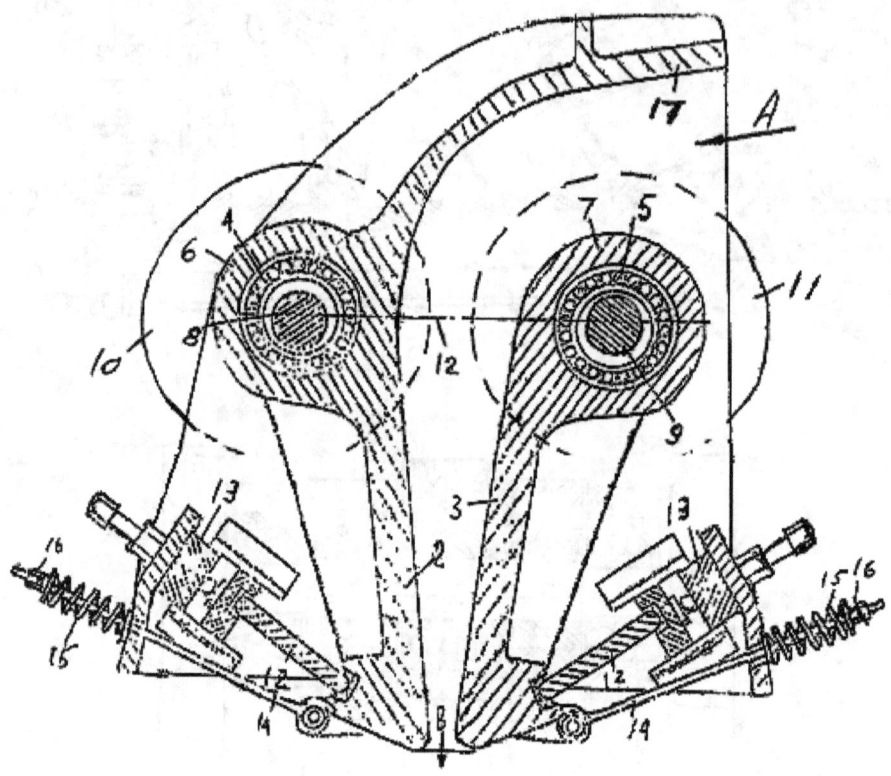

Since the present crusher utilizes the gyratory movement of the hinge, it can be compared with known gyratory crushers and eccentric-roll crushers (e.g. of Rapidex Inc. and Joy Manufacturing Co.). The working stroke of the present crusher is almost doubled in comparison with conventional jaw crushers. Therefore, it works more uniformly and lessens the required horsepower, dynamic load and mass. The developed torsional oscillations are two times less dangerous to its foundation, than rotational oscillations developed at the opposite rotation of the eccentrics in twin-jaw crushers. However, the jaws of the latter crushers serve longer. Inherent in single-jaw crushers, this disadvantage is tolerable, especially for non-abrasive rocks, and the jaws rubbing enhances crushing.

IN THE DRAWING

A lone figure is a cross-sectional view of the present crusher.

DESCRIPTION OF THE PREFERRED EMBODIMENT

The present crusher consists of a frame 1 to which swing jaws 2 and 3 are mounted in bearings 4 and 5 of their hinges 6 and 7. The latter are installed on eccentric shafts 8 and 9 in bearing supports (not shown) of the frame 1 and have mechanical drives 10 and 11. The eccentrics are directed oppositely from the line 12 connecting their centers and rotate in the same direction.

The lower ends of the jaws 2 and 3 are releasable and fixed in position by toggle arrangements 12. The latter are safety mechanisms and have a series of shims 13 used to adjust the position of the jaws 2 and 3 to vary the discharge opening between them. The toggle arrangements 12 are held in place by tension rods 14 mounted on the frame 1. The rods have springs 15 and threaded ends carrying nuts 16 and extended through suitable holes in the frame 1.

The upper part of the jaw 2 has an arc extension 17 wrapped around the interacting eccentric hinge 7 of the opposite jaw 3 and forming a primary crushing chamber in the converging passage between the arc and hinge.

Rocks are fed into the top of the converging passage, as indicated with arrow A, and are moved downward into the space between the jaws 2 and 3. The material trapped in the converging passage is crushed due to the vertical component of the jaws movement. At this, a peristaltic pumping action enhances throughput of the rocks through the crusher. The material trapped between the jaws is crushed by the horizontal component of their movement (as in conventional jaw crushers) and is discharged through the opening between the jaws 2 and 3, as indicated with arrow B.

Such a crusher construction improves the conditions of its exploitation, decreases the dynamic forces, loads its mechanical drive more uniformly, increases the output and decreases the mass and requirements to the foundation.

I.15. TWIN-JAW CRUSHERS

BACKGROUND OF THE INVENTION

This invention relates to overhead eccentric jaw crushers with both the jaws movable. Such known twin crushers (e.g. manufactured by Iowa Manufacturing Co.) consist of two synchronized swing jaws mounted directly on eccentric shafts to get the downward-and-forward motion. The lower ends of the jaws are held in position against the toggles by tension rods. In comparison with single-jaw crushers, the twin-jaw crushers are better balanced, have 40% greater capacity and 2 - 4 times longer service life of their jaws (foreign data).

The objective of the present invention is to elevate the efficiency of the above crushers still more, simplify the construction and decrease the mass and requirements to the strength, balancing and mechanical drive.

SUMMARY OF THE INVENTION

Above objective is achieved by providing the swing jaws with ears hinged on the shaft eccentrics situating 180° out of the phase. In this way, the moving masses are almost completely counter-balanced and do not subject the crusher's frame and foundation to vertical oscillations (in contrast to the known crushers).

However, the jaws of the present invention will have less service life than those of twin crushers because they move in opposite vertical directions. Being inherent in single jaw crushers, this disadvantage is tolerable, especially for non-abrasive rocks, the rubbing jaws enhancing the crushing.

The first swing jaw (with ears) can be also provided with an arc extension wrapped around the interacting eccentric hinge of the opposite swing jaw and forming a primary crushing chamber in the converging passage between the arc and the hinge. At this, the vertical component of the jaw's working stroke is also utilized, whereas at present only the horizontal component of the latter is used. This almost doubles the working displacement and the crushing area, the mechanical drive working more uniformly, with the lesser horsepower, dynamic load and mass.

IN THE DRAWINGS

Fig. 1 is a cross-sectional view of the first embodiment of the present invention.
Fig. 2 is the same as above, the second embodiment.

Fig. 3 is a cross-sectional view along lines III - III in Fig. 1 and 2.

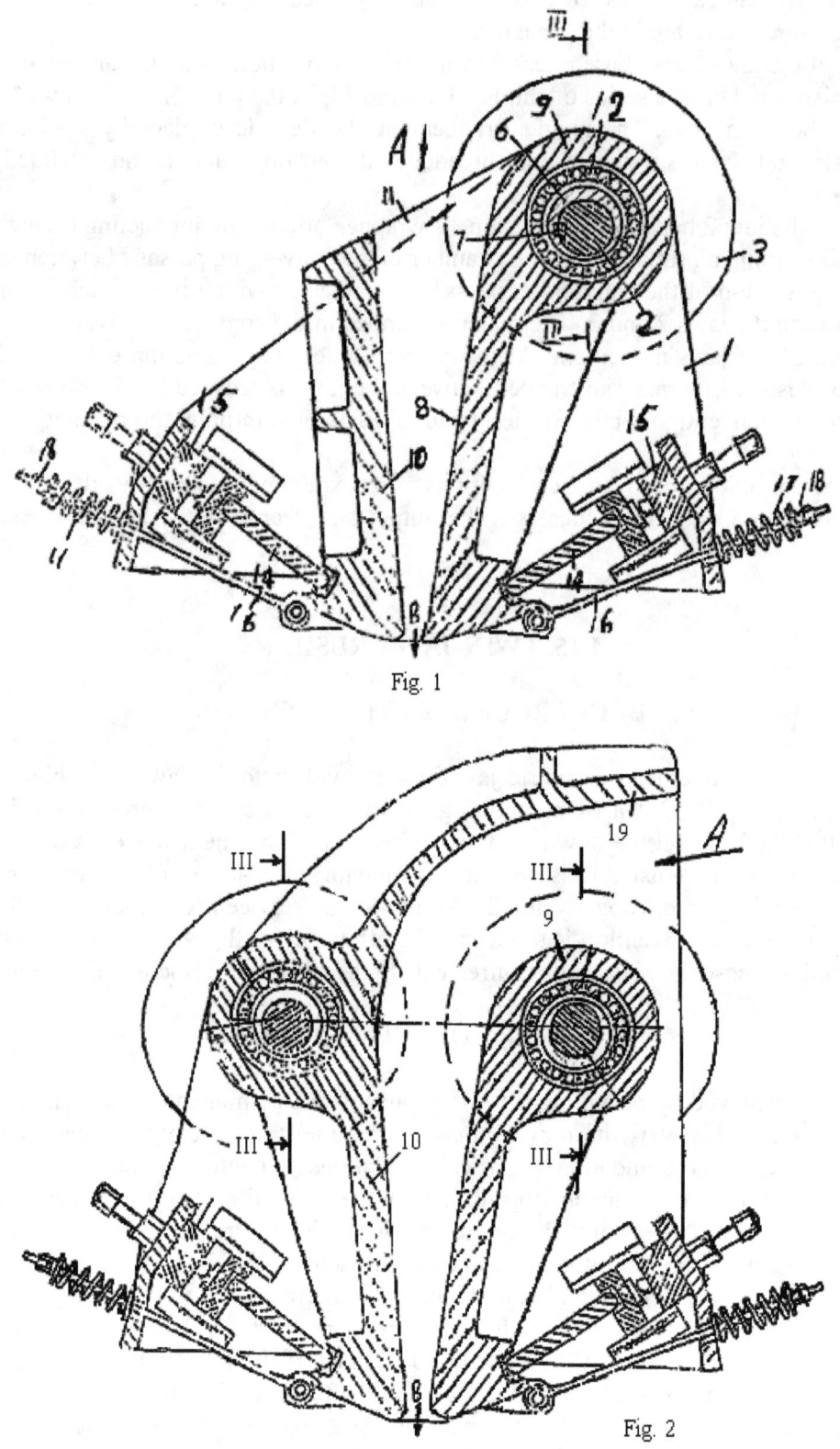

Fig. 1

Fig. 2

DESCRIPTION OF THE PREFERRED EMBODIMENTS

The present crusher consists of a frame 1 on which a shaft 2 with a mechanical drive 3 is mounted in bearing supports 4. To the shaft 2, eccentric collars 5 and 6 are fixed by a key or a similar joint 7, 180° out of phase. One swing jaw 8 is mounted directly with its hinge 9 on the collars 6, and another jaw 10 by ears 11 on the collars 5 in bearings 12 and 13.

The lower ends of the jaws 8 and 10 are releasable and fixed in position by toggle arrangements 14. The latter are safety mechanisms and provided with a series of shims 15 that easily adjust the position of the jaws 8 and 10 to vary the discharge opening between them. The toggle arrangements 14 are held in place by tension rods 16 mounted on the frame 1 and have springs 17 and threaded ends, which carry nuts 18 and extend through suitable holes in the frame 1.

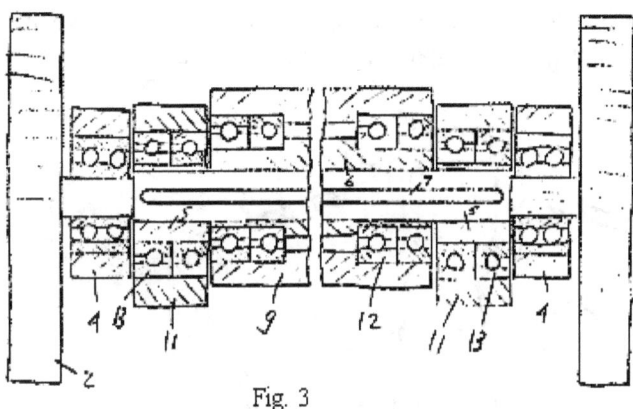

Fig. 3

The upper part of the jaw 10 in Fig. 2 is provided with an arc extension 19 wrapped around the interacting eccentric hinge 9 of the opposite jaw 8 and forming a primary crushing chamber in the converging passage between the arc and hinge.

Rocks are fed into the crusher as indicated with arrow A, crushed by the horizontal component of jaws movement and discharged through the opening between them as indicated with arrow B. In the second embodiment (Fig. 2), the material trapped in the converging passage is crushed due to the vertical component of the jaws movement. At this, a peristaltic pumping action enhances throughput of the material through the crusher.

II FLUID-DRIVEN ROCK CRUSHERS

II.1. HYDRAULIC CONE CRUSHER
BACKGROUND OF THE INVENTION

This invention relates to cone rock crushers. At least two their crushing members have material engaging curved crushing surfaces and are located with respect to each other such that a passageway of progressively narrow width is formed along said material engaging surfaces. Their drive means operationally connected with at least one said crushing member for imparting a movement away and toward another crushing member.

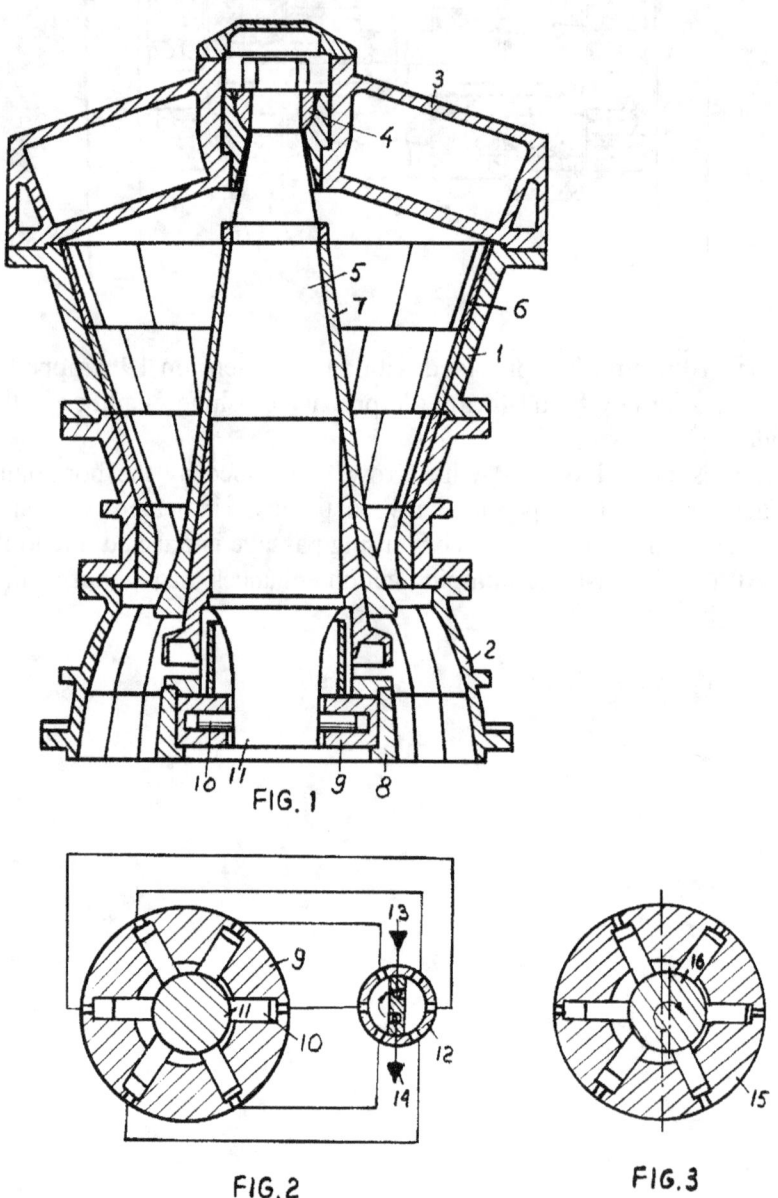

FIG. 1

FIG.2   FIG.3

The drive means of such known crushers (e.g. US patents Nos. 3,506,204; 3,754,716; 3,850,376; 3,873,057; 4,168,036; 4,192,472 and many others) comprises a rotating eccentric with a clumsy, heavy mechanical drive.

The objective of the present invention is to eliminate this disadvantage, simplify the construction, decrease the down-time and achieve still more technical and economical effects which are more apparent and listed below in the Summary of the Invention.

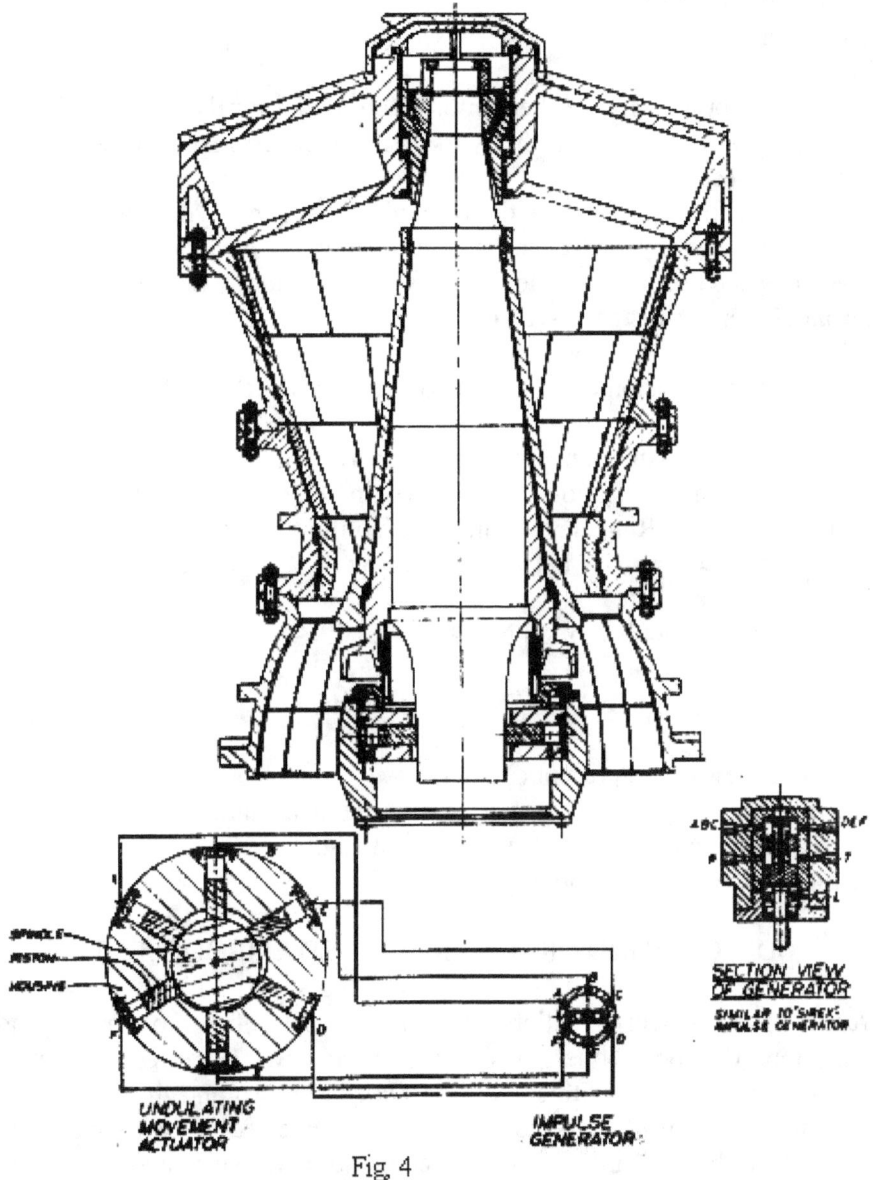

Fig. 4

SUMMARY OF THE INVENTION

Above objectives are achieved by that said drive means comprises several expansion fluidic chambers situated radially with respect to the curved surface of said driven crushing member and connected with a source of pressurized fluid one by one in such a manner that a progressive radial undulating movement along said passageway is created progressively from one said chamber to another. Thus, the merit of the present invention lies in the use of a hydrostatic drive in place of conventional drive trains. For gyratory and cone crushers, for example, this offers potential for 30% height reduction, a better crushing performance for given space and metallic mass and the size reduction of the auxiliary equipment and of the premises.

Said hydrostatic drive is not a hydraulic version of an electrical or diesel drive, since there is no rotary movement of the driven shaft. The usual advantages of hydrostatics are also exploited:

1) the direct connection to the crusher as a shaft mounted unit;
2) the accurately limited full starting torque;
3) the instantaneous non-destructive load limitation;
4) the speed control, reversibility (for releasing material) and feed back control;
5) the high reliability;
6) the limited maintenance requirement;

7) the prime mover can be started unloaded;

8) the low cost height and weight;

9) the simple installation.

This invention is attractive not only for crushers in production, but for the modification of those in operation as well. This greatly expands the market, all the work being reduced mainly to assembly operations almost without significant capital investments.

This invention can be also made in line with the author's earlier hydraulic shock/impact applications. Therein an impulse exciter/pulser of hydraulic shocks is installed in the pipeline with its length greater than a half product of the sound velocity and the impulse time. Please see Addendum with the following publications:

S.I. Fishgal, The Calculation of Pistonless Hydroshock Devices, Soviet Applied Mechanics, Consultants Bureau, New York, June, 1975

S.I. Fishgal. Re Computation of Pistonless Hydroimpact Devices. Applied Mechanics, v. 9, Kiev, 1973, No. 11 (in Russian)

S.I. Fishgal. Hydraulic Impact Nut-Runners. Machines & Tooling. Vol. XLI, London, 1970, No. 1. Production Engineering Research Association of Great Britain

S.I. Fishgal. Hydraulic Shock Nut-Runners. Machines & Tooling. Moscow, 1970, No. 1

S.I. Fishgal. Hydraulic Impact Drive of a Grab-Bucket Shell. Construction and Road-Making Machines. Moscow, 1968, No. 10 (in Russian)

S.I. Fishgal. Hydraulic Grab. Soviet Invention No. 449,870; 1964

IN THE DRAWING

Fig. 1 is a cross-sectional view of the present crusher,

Fig. 2 is a hydraulic diagram of a drive means with a rotating distributor.

Fig. 3 is a diagram of a fluidic pulser of the drive means of Fig. 1.

Fig. 4 is a more detailed drawing corresponding to Fig. 1-3.

DESCRIPTION OF THE PREFERRED EMBODIMENT

In the gyratory crusher of the present invention (FIG. 1), a bowl 1 is fixed to a base 2 and a cross-member 3 in which a spherical support 4 for a cone 5 is inserted. The bowl 1 and cone 5 represent said crushing members and are provided accordingly with smooth or ribbed plates 6 and liners 7 of manganese steel. The base 2 has a housing 8 for a drive means 9. The housing 8 is connected to the base 2 by one or more rib-like details (not shown). The described features are close to conventional designs and are self-explanatory.

The drive means 9 comprises hydraulic cylinders 10 (representing said fluidic chambers) mounted radially around a vertical axle 11 of the cone 5 and resembles a stator of a hydraulic, radial-piston motor or a pump without built-in rotating distributor and with the axle 11 in place of a conventional shaft.

Fluid supply of the drive means 9 is achieved either from a rotating fluidic distributor 12 (Fig. 2) or a valve (not shown) connected to a source of a pressurized fluid 13 and a tank 14, or from a pulser 15 (Fig. 3) with the same quantity of fluidic chambers.

In the second case, the pulser 15 represents the same stator of the drive means 9, the difference lying in providing an eccentric shaft (rotor) 16 in place of the axle 11. The hydraulic cylinders of the drive means 9 and the pulser 15 are interconnected accordingly with each other in such a manner that when the eccentric shaft 16 rotates and creates radial undulating movement in the cylinders of the pulser, the connected cylinders of the drive means reproduce the same movement (said hydraulic connections are not shown).

Such a hydraulic drive is very simple and efficient. However, since the drive means and pulser are of the same order, not full economy in metallic mass is achieved for relatively slow applications.

In the latter case, when the revolutions of the pulser shaft are much less than those of a prime mover (e.g. an electrical motor or internal combustion engine), the most economical means is a conventional pump unit with the rotating distributor or the valve. In this case, the revolutions of the pump are the same or of the same order of the mover, and gearing, not necessarily mechanical, is provided only for the distributor or the valve (an electromagnetic valve can be also used). Therefore, if to compare this drive with conventional mechanical drives of known crushers wherein the gearing is provided for the full drive, in the present case the gearing is provided only for the relatively small distributor which can be considered here as a fluidic amplifier. Thus, the clumsiness of the mechanical drive is eliminated.

The described undulating movement of the drive means 9 causes the cone 5 swings about the vertical axis of the crusher. When the cone approaches the bowl, the rock material fed past the cross-member 3 is crushed and falls down under the force of gravity when the cone moves away. This process continues until the stone material is reduced to a size small enough to pass through the discharge opening.

It is obvious that many modifications and adaptations can be made without departing from the spirit and scope of the invention.

II.2. HYDRAULIC GYRATORY CRUSHER

BACKGROUND OF THE INVENTION

This invention relates to gyratory rock crushers including a bowl and a cone driven alternately toward and away from each other. In such known gyratory crushers, an eccentric shaft or an unbalanced rotor drives the cone. The objective of the present invention is to decrease the mass and dimensions of the crusher, to simplify its construction and reduce the drive to easily accessible and interchangeable conventional relatively cheap units.

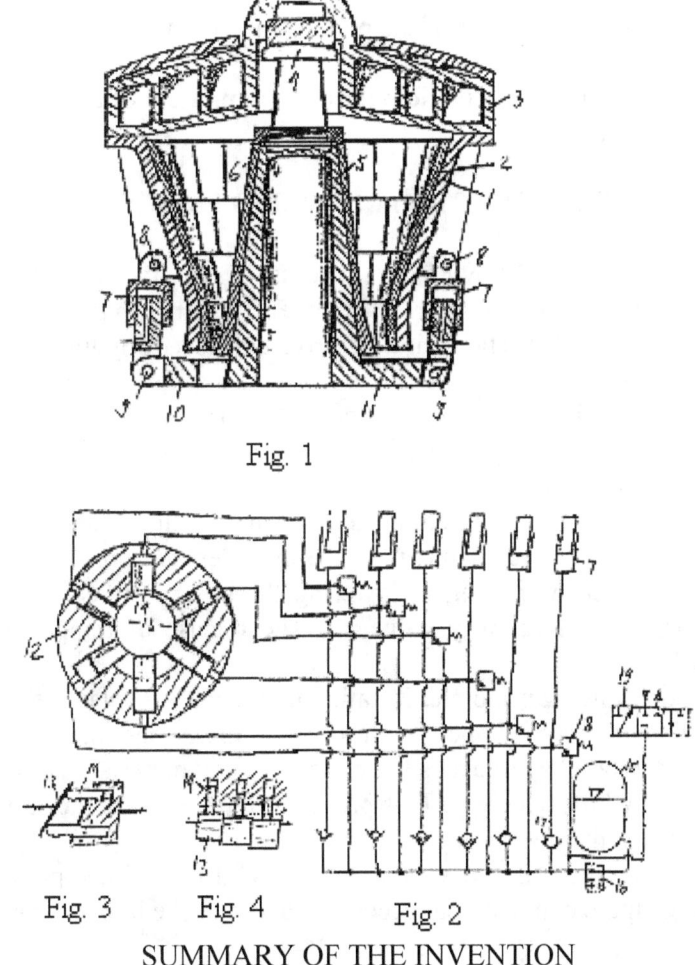

Fig. 1

Fig. 3   Fig. 4   Fig. 2

SUMMARY OF THE INVENTION

The above objective is achieved by executing the drive as hydraulic cylinders that act on a swing-working member, are spaced equally apart from each other on the circumference of the crusher and connected to a hydraulic pulser. The latter represents a system of the same quantity of hydraulic cylinders with their plungers actuated by a displacer such as radial or axial cams, eccentrics and the like. This simplifies the drive by eliminating heavy components of the known mechanical drives and hydraulic units with lesser masses and cost. Besides, the safety mechanism for releasing non-breakable materials and controlling the opening operates easier.

Similar to the hydraulic cone crusher of the previous chapter, this invention can be also made in line with the author's earlier hydraulic shock/impact applications. Therein an impulse exciter/pulser of hydraulic shocks is installed in the pipeline with its length greater than a half product of the sound velocity and the impulse time. Please see Addendum with the following publications:

S.I. Fishgal, The Calculation of Pistonless Hydroshock Devices, Soviet Applied Mechanics, Consultants Bureau, New York, June, 1975

S.I. Fishgal. Re Computation of Pistonless Hydroimpact Devices. Applied Mechanics, v. 9, Kiev, 1973, No. 11 (in Russian)

S.I. Fishgal. Hydraulic Impact Nut-Runners. Machines & Tooling. Vol. XLI, London, 1970, No. 1. Production Engineering Research Association of Great Britain

S.I. Fishgal. Hydraulic Shock Nut-Runners. Machines & Tooling. Moscow, 1970, No. 1

S.I. Fishgal. Hydraulic Impact Drive of a Grab-Bucket Shell. Construction and Road-Making Machines. Moscow, 1968, No. 10 (in Russian)

S.I. Fishgal. Hydraulic Grab. Soviet Invention No. 449,870; 1964

IN THE DRAWINGS

Fig. 1 is a cross-sectional view of the crusher of the present invention.

Fig. 2 is its hydraulic diagram.

Fig. 3 and 4 kinematic diagrams of the hydraulic pulsers appropriately of a radial eccentric (cam) type with parallel plungers, and of an axial cam type.

DESCRIPTION OF THE PREFERRED EMBODIMENTS

The present hydraulic crusher contains a bowl 1 with its crushing cavity lined with smooth or ribbed plates 2 of manganese steel. The bowl 1 is fixed with a cross-member 3 resting on a spherical support 4 of a cone 5. The latter is also provided with liners (6). The hydraulic drive consists of hydraulic cylinders 7 swung in axles 8 and 9 mounted on the circumference of the external surface of the bowl 1, and a ring plate 10 joined to the cone 5 with several ribs 11 and serving as a base for the crusher. The hydraulic cylinders 7 are connected to a multi-plunger hydraulic pulser 12 consisting of a rotating member 13, such as a radial eccentric (Fig. 2 and 3) or axial cam (Fig. 4) acting on plungers 14.

For leakage replenishment, the cylinders are connected to a hydraulic accumulator 15 via a two-way valve 16 and non-return valves 17. The pressure chambers of the cylinders are provided with safety valves 18. The discharge opening width is adjusted by raising or lowering the bowl 1 by connecting hydraulic cylinders 7 with pressure and drain lines via a three-way valve 19. The release of trapped non-breakable material is achieved by raising the bowl 1.

The crusher operates on the following principle. Rotating member 13 displaces the hydraulic liquid from the pulser cylinders into the cylinders 7 one by one. This raises the appropriate part of the bowl 1 and it is swung about its vertical axis. While the bowl approaches the cone, the rock material is crushed and falls down under the force of gravity when the bowl moves away. This process continues until the stone material is reduced to a size small enough to pass through the discharge opening.

Since the pulser serves as an intensifier too, the hydraulic drive of the present crusher is much less cumbersome in comparison with the conventional eccentric drives of the known crushers.

II.3. HYDRAULIC JAW CRUSHERS

BACKGROUND OF THE INVENTION

This invention relates to jaw rock crushers. In such known crushers, eccentrics or unbalanced rotors drive one or two oppositely situated swing jaws. The disadvantage of the known jaw crushers in comparison with known gyratory crushers lies in their cyclical work, i.e. the jaw makes one power stroke and one idle stroke during one revolution of the eccentric shaft. That requires higher strength of construction, balancing and a more powerful mechanical drive. Also, in the conventional jaw crushers, the axle of the swing jaw is situated in its rear part, the power stroke of the jaw providing a force component acting in anti-gravitational direction and braking the material.

The objectives of the present invention is to increase the efficiency of above crushers, decrease their mass and requirements to the construction strength, balancing and mechanical drive, to simplify the construction, and to reduce the drive to easily accessible and interchangeable conventional relatively cheap units.

SUMMARY OF THE INVENTION

Above objectives are achieved by executing the drive in form of hydraulic cylinder acting on the swing jaw and connected to a hydraulic pulser. The latter is made as a hydraulic reciprocating cylinder with its plunger actuated by a displacer, such as radial or axial cam, eccentrics, cranks and the like. This simplifies the drive by eliminating heavy components of the known mechanical drives and reduces it to easily accessible and interchangeable hydraulic units with lesser masses and cost.

Above cyclical work is eliminated by dividing the swing jaw into separate sections interacting with hydraulic cylinders connected to a multi-plunger hydraulic pulser. The plungers of the latter are spaced equally apart from each other and relatively to the displacer.

Such a crusher squeezes treated materials continuously since the jaw sections work in turn, the more sections, the more uniformly the crusher works. The different jaw sections squeeze and release the materials simultaneously. Such uniform work lessens the horsepower, dynamic load and the crusher's mass and foundation. Besides, the efficiency of the breaking down of rocks is elevated since the transverse (bending) load on rocks between the different sections is probable.

Still better balancing is achieved providing opposite swings of the synchronized swing jaws, instead of the stationary one. Besides, the opposite swing counterbalanced jaws increase the crusher's throughput.

To facilitate the gravitational movement of the material and to provide the possibility of a much lower headroom construction (e.g. for portable crushers of underground mining), the jaws form an inclined crushing chamber. It slops downwardly from the inlet to the discharge, with the swing jaw being lower and having frontal ears for its swing axle. Such a low-profiling crusher with its dramatically reduced mass and easily disconnected hydraulic drive makes it suitable for underground mine applications.

Similar to the hydraulic cone and gyratory crushers of the previous chapter, this invention can be also made in line with the author's earlier hydraulic shock/impact applications. Therein an impulse exciter/pulser of hydraulic shocks is installed in the pipeline with its length greater than a half product of the sound velocity and the impulse time. Please see Addendum with the following publications:

S.I. Fishgal, The Calculation of Pistonless Hydroshock Devices, Soviet Applied Mechanics, Consultants Bureau, New York, June, 1975

S.I. Fishgal. Re Computation of Pistonless Hydroimpact Devices. Applied Mechanics, v. 9, Kiev, 1973, No. 11 (in Russian)

S.I. Fishgal. Hydraulic Impact Nut-Runners. Machines & Tooling. Vol. XLI, London, 1970, No. 1. Production Engineering Research Association of Great Britain

S.I. Fishgal. Hydraulic Shock Nut-Runners. Machines & Tooling. Moscow, 1970, No. 1

S.I. Fishgal. Hydraulic Impact Drive of a Grab-Bucket Shell. Construction and Road-Making Machines. Moscow, 1968, No. 10 (in Russian)

S.I. Fishgal. Hydraulic Grab. Soviet Invention No. 449,870; 1964

IN THE DRAWINGS

Fig. 1 is a side elevation view of the present crusher with a stationary jaw and a double-acting hydraulic cylinder.

Fig. 2 is a top plan view of the crusher with two-sectional swing jaws.

Fig. 3 is a hydraulic diagram of the crusher of Fig. 2.

Fig. 4 is a kinematic diagram of a hydraulic pulser of a crank type.

Fig. 5 is the same as above of a radial eccentric (cam) type with parallel plungers.

Fig. 6 is the same as above of an axial cam type.

Fig. 7 is the same as in Fig. 1, with a single-acting hydraulic cylinder.

Fig. 8 is above crusher's hydraulic diagram.

Fig. 9 is the same of the crusher of Fig. 1 with two opposite synchronized swing jaws.

Fig. 10 is the same as above, a swing jaw having frontal ears.

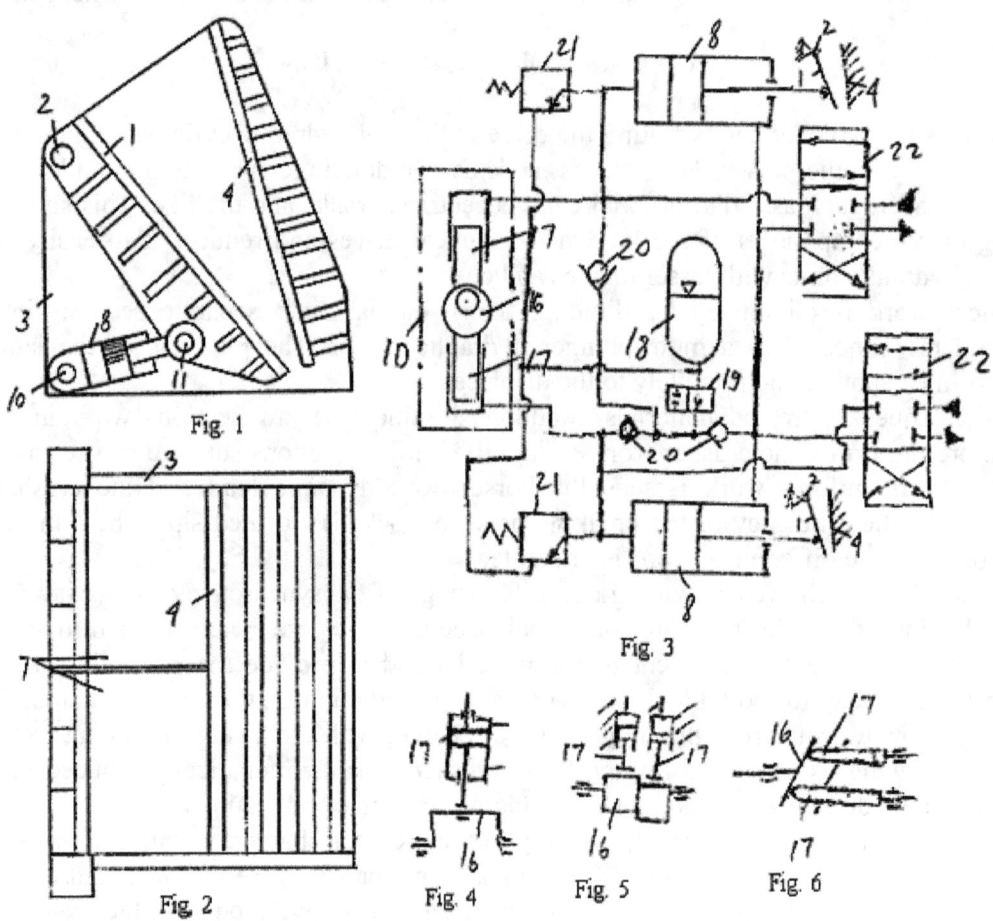

Fig. 1

Fig. 2

Fig. 3

Fig. 4

Fig. 5

Fig. 6

DESCRIPTION OF THE PREFERRED EMBODIMENTS

A hydraulic jaw crusher of the present invention consists of a swing jaw 1 swung on an axle 2 mounted on a frame 3 (Fig. 1 and 2). An opposite jaw can be a stationary one 4 fixed to the frame 3 (Fig. 1 - 3, 7 and 8) or a synchronized jaw 5 swun on an axle 6 mounted either on the frame 3 (Fig. 9) or on the same axle 2 (Fig. 10). To eliminate the cyclical work of the crusher, the swing jaw 1 is divided into separate sections 7 (two such sections are shown in Fig. 2 only in way of illustration, but not in a limiting sense). The hydraulic drive consists of hydraulic cylinders 8 swung in axles 9 mounted on the frame 3 and acting upon the swing sections. The cylinders of different sections are connected to a multi-plunger hydraulic pulser 10.

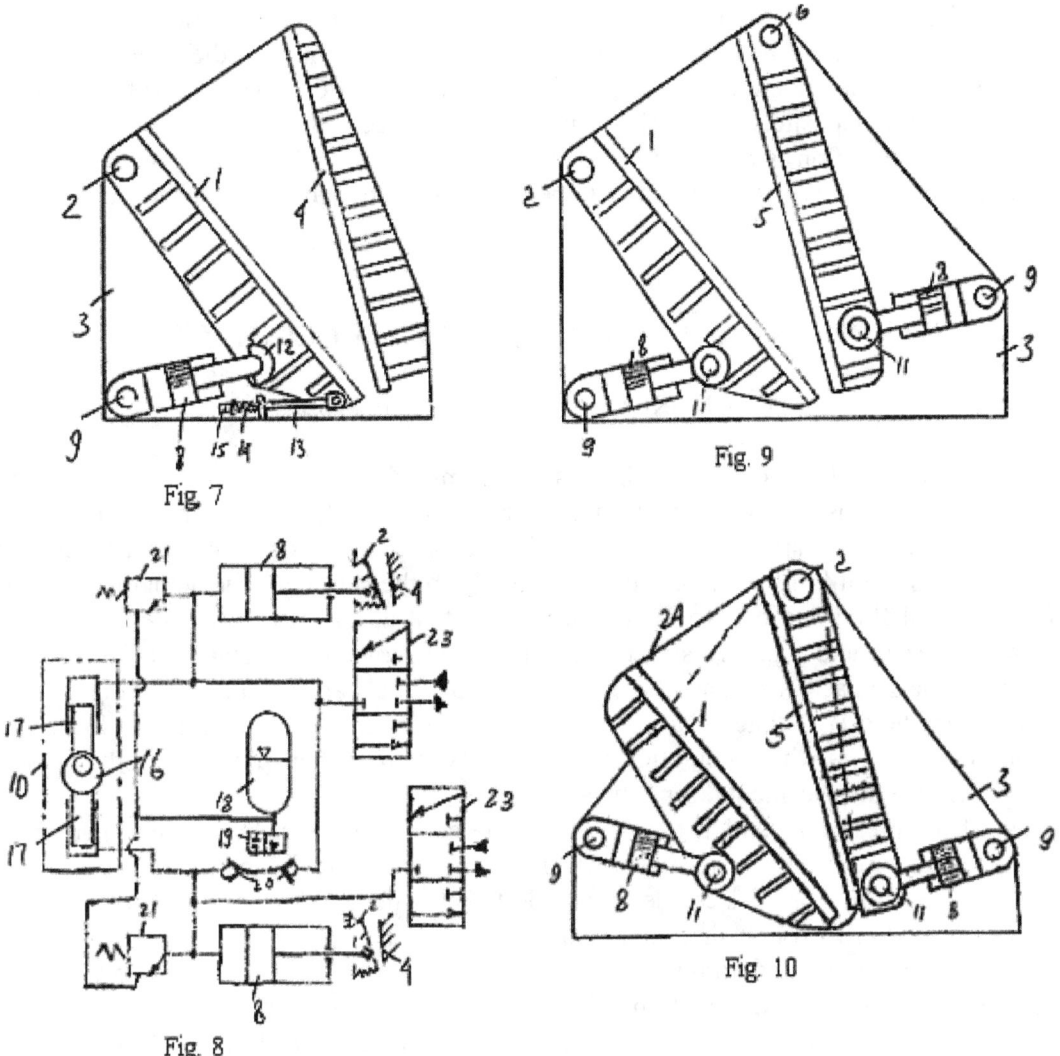

Fig. 7

Fig. 9

Fig. 8

Fig. 10

In double-acting cylinders (Fig. 1, 3, 9 and 10), their rods are hinged to the sections with axes 11. In single-acting cylinders (Fig. 7 and 8), their rods are provided with spherical supports 12 fixed to the sections, although hinges can be also used as above. With single-acting cylinders, tension rods 13 are mounted on the frame 3 and have threaded ends carrying springs 14 and nuts 15. Said ends extend through suitable holes in the frame 3. The springs 14 provide the backstroke. Such a tension arrangement holds a single-acting cylinder with an unhinged rod in place. The hydraulic pulser 10 consists of a rotating member 16, such as a radial eccentric (Fig. 3, 5 and 8) or a cam (not shown), a crank, an axial cam, (Fig. 6) and the like, acting on plungers 17.

To replenish leakages, the cylinders are connected to a hydraulic accumulator 18 via a two-way valve 19 and non-return valves 20. The pressure chambers of the cylinders have safety valves 21. To adjust the positions of jaws, (to vary the discharge opening between them) and to release trapped unbreakable materials, the cylinders are connected to pressure and drain lines by four-way valves 22 (Fig. 3) or three-way valves 23 (Fig. 8).

In the crushers with both swing jaws Fig. 9 arid 10), the cylinders of the opposite sections have parallel connections to the pulser 10. To provide a downward component of the power stroke, the lower swing jaw has frontal ears 24 for its swing axle (Fig. 10).

Rocks are fed from the top, crushed by the jaw sections moving to the opposite jaw and fall down when the sections are moved back. The rocks are crushed again by the next power stroke until they drop through the discharge opening. The resistance of the motion of the moving section increases sharply on the power stroke and diminishes on the return one. That is why the sections of the present crusher distribute the load on the mechanical drive of the pulser, frame and foundation more uniformly and ensure the smooth running. For assistance to the distribution, a hydraulic accumulator can be used in a conventional manner (not shown), storing the surplus energy during the idle stroke and releasing it for crushing on the power stroke.

Big rocks are subjected to transverse (bending) load between sections. That elevates the efficiency still more. The crusher with oppositely swing jaws is counterbalanced almost completely and, therefore, its dynamic load on the foundation is considerably decreased. Since the pulser serves as an intensifier too, the hydraulic drive of the present crusher is not as cumbersome as the conventional eccentric drive of the known crushers.

II.4. FLUID-ACTUATED GYRATORY CRUSHER

BACKGROUND OF THE INVENTION

This invention relates to fluid-actuated rock crushers comprising at least two crushing members having material engaging curved crushing surfaces. With the respect to each other, the members are situated such that a passageway of progressively narrow width is formed along said surfaces and moved alternatively away and toward each other by several fluid actuators situated radially to the surfaces. The actuators are connected to a source of a pressurized fluid one by one in such a manner that at least two opposite actuators respectively expand and are squeezed, and a progressive radial undulating movement along said passageway is created progressively from one said actuator to another in response to said expansion and squeezing.

In such known crushers (US Patent No. 2,291,910 and German Patent No. 1,157,459), said actuators constitute hydraulic cylinders with a complicated force transfer from their plungers to the crushing member (because of their relative displacement during the stroke).

The objective of the present invention is to simplify the construction, decrease the downtime, height and mass and to achieve better crushing performance for given space and metallic mass.

The usual advantages of hydrostatics can be also exploited as follows:

direct connection to the crusher as a shaft mounted unit;

full starting, but accurately limited, torque;

instantaneous, nondestructive load limitation;

Speed control, reversibility (for releasing non-breakable material) and feed back control;

high reliability and limited maintenance requirements;

prime mover can be started unloaded;

low cost and simple installation.

This greatly expands the market. Besides, this invention is attractive not only for crushers in production, but for the modification of those in operation as well, all the work being reduced mainly to assembly operations without significant capital costs.

SUMMARY OF THE INVENTION

Above objectives are achieved by that said actuators constitute chambers of a resilient material. This invention provides the appropriate size reduction of the auxiliary equipment of the crushing plants and more reduction of the premises.

IN THE DRAWINGS

FIG. 1 is a cross-sectional view of the present crusher;
FIG. 2 is a hydraulic diagram of a drive means with a rotating distributor.

DESCRIPTION OF THE PREFERRED EMBODIMENT

In the gyratory-like crusher of the present invention (FIG. 1) a bowl 1 is fixed with a base 2 and a cross-member 3 in which a spherical support 4 for a cone 5 is inserted. The bowl 1 and cone 5 represent said crushing members and have accordingly smooth or ribbed plates 6 and liners 7 of manganese steel. The base 2 has a housing 8 for fluidic chambers 9 of a resilient material. The housing 8 is connected to the base 2 by one or more rib-like details (not shown). These features are close to conventional designs and are self-explanatory. The chambers 9 are mounted radially around a vertical axle 10 of the cone 5.

Fluid supply of the chambers 9 is achieved either from a rotating fluidic distributor 12 (FIG. 2) or a valve (not shown) connected to a source of pressurized fluid 13 and a tank 14.

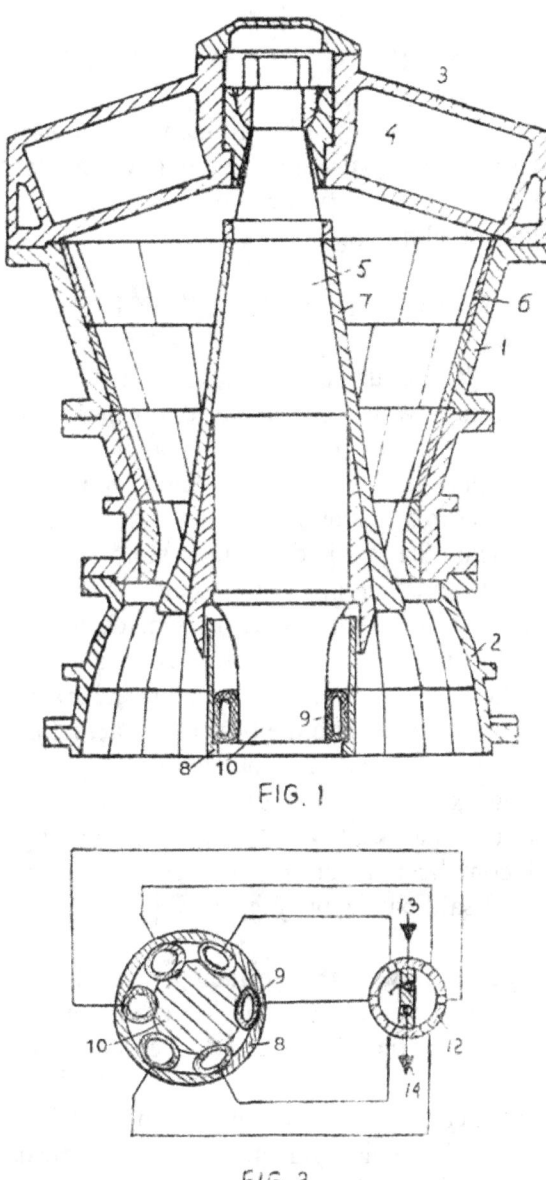

FIG. 1

FIG. 2

In such a hydraulic drive, the revolutions of a pump are the same or of the same order of the mover, and gearing, not necessarily mechanical, is provided only for the distributor or valve (an electromagnetic valve can be also used). Therefore, if to compare this drive with conventional mechanical drives (wherein gearing is provided for the full drive), in the present case, gearing is provided only for the relatively small distributor which can be considered here as a fluidic amplifier. Thus, the clumsiness of the mechanical drive is eliminated.

The chambers 9 are fed one by one in such a manner that at least two opposite chambers respectively expand and are squeezed, a progressive undulating movement along the passageway between the bowl and cone being created progressively from one chamber to another in response to their expansion and squeezing.

The undulating movement of the chambers 9 swings the cone 5 about the vertical axis of the crusher. When the cone approaches the bowl, the rock material fed past the cross-member 3 is crushed and falls down under the force of gravity after the cone moves away. This process continues until the stone material is reduced to a size small enough to pass through the discharge opening.

It is obvious that many modifications and adaptations can be made without departing from the spirit and scope of the invention.

II.5. FLUID-ACTUATED JAW CRUSHERS

BACKGROUND OF THE INVENTION

This invention relates to fluid-actuated jaw crushers. Such known crushers comprise a frame with a pivotally mounted swing jaw provided with a hydraulic drive (e.g. U.S. Patent Nos. 2,620,629 and 3,386,667). The objective of the present invention is to simplify the construction, decrease its mass with the possibility of making a portable crusher for underground mining.

SUMMARY OF THE INVENTION

Above objective is achieved with a fluid-actuated drive having an elastic closed-contour member connected to a pressurized-fluid source and sealing a working chamber defined by a frame base and a swing jaw. If the ends of said member are sealed during all the working cycle (as in known diaphragm, membrane, bellows, rolling sleeve and the like actuators), said source provides pulsating pressurized fluid.

Since the working area of the described actuator is much more than that of hydrocylinders of known hydraulic crushers, the working stress on the jaw is distributed more evenly. That is why the mass of the jaw and its vibration is less damaging for the crusher and its foundation. This decreases the mass of the crusher.

Still further improvement is achieved when said actuator is made to work in self-oscillating regime and connected to a static fluid-pressure source, such as a compressed air main. To get the self-oscillations, a passageway to the surrounding atmosphere is provided in the collar near an elastic sealing lip of said member. Since the jaw swings, the angle between the jaw and the base of the actuator increases during a power stroke.

In the initial position, the lip can be tightly pressed to the base or the jaw. At the end of the power stroke, the portion of the lip close to the swing axis can be still sealed, but the portion close to the opposite end will define a passageway for the working chamber to lead to the surrounding atmosphere. This simplifies the construction still more since no complicated control devices are necessary (unlike in U.S. Patent No. 3,386,667). To secure the forming of said passageway in the determined zone, a limiter of the deflection of the lip can be provided.

In another variant of the collar mounting with the lip facing either jaw, or frame base, the collar is fixed accordingly either to the base, or to the jaw. In still another modification, the passageway opening to the surrounding atmosphere has one or more orifices in the elastic member, said orifices being sealed in the initial position and opened at the end of the power stroke.

For balancing of the crusher and increasing its throughput, an opposite synchronized swing jaw can be provided instead of the stationary jaw. To avoid cyclical work of the crusher, the swing jaw can be divided into several separate sections provided with working chambers acting in opposite directions.

IN THE DRAWINGS

Fig. 1 is a cross-sectional view of the present crusher with one swing jaw.
Fig. 2 is a top view of the above crusher with a two-section swing jaw.
Fig. 3 is the same as in Fig. I, with two oppositely situated synchronized swing jaws.
Fig. 4 is an elevation cross- sectional view of a bellows actuator.
Fig. 5 is a perspective view featuring a collar deflection limiter.
Fig. 6 is an elevation cross-sectional view featuring a flat collar fixed to the base.
Fig. 7 is the same as above with a flat collar fixed to the jaw.
Fig. 8 is the same as above with a collar having outlet openings on its lips.

Fig. 9 is an elevation cross-sectional view featuring a diaphragm with outlet openings on its lateral surface.

DESCRIPTION OF PREFERRED EMBODIMENTS

A jaw crusher of the present invention consists of a swing jaw 1 on an axle 2 mounted on a frame 3 (Fig. 1 - 3). An opposite jaw can be a stationary one 4 fixed to the frame 3 (Fig. 1 and 2) or another synchronized swing one 1 on another axle 2 mounted on the frame 3 (Fig. 3). To eliminate the crusher cyclical work, the swing jaw can be divided into separate sections 7 (two such sections are shown in Fig. 2 only in way of illustration, but not in a limiting sense). The fluid actuator consists of an elastic closed-contour member 8 positioned between a base 9 of a frame 3 and the swing jaw. The actuator is capable to seal a working chamber connected to a pressurized fluid source (not shown).

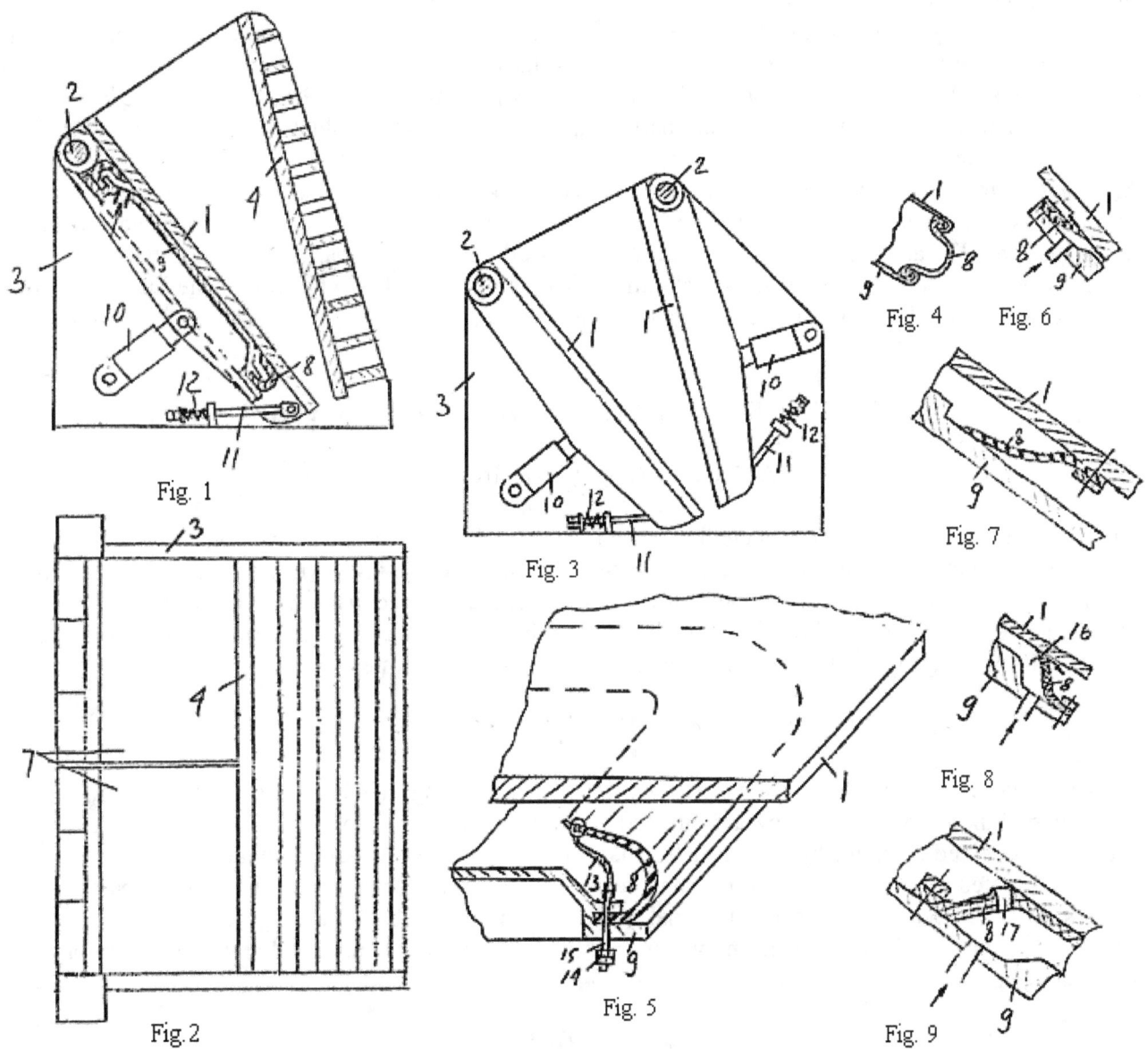

Fig. 1
Fig. 2
Fig. 3
Fig. 4
Fig. 6
Fig. 7
Fig. 8
Fig. 5
Fig. 9

The base 9 swings on the axle 2 and is held in place by a hydrocylinder 10 hinged between the base 9 and the frame 3. The swing jaw is held in place by its weight and can be assisted by a rod 11 with a spring 12. Such a construction allows the release of unbreakable material and permits changing the opening of the crusher. These features are close to conventional designs and are self-explanatory.

Depending on the desirable work regime, the working chamber can be connected to a source of pulsing or constant pressurized fluid.

In the first case, the ends of the pliable elastic member 8 are sealed during all the working cycle as in known diaphragm (Fig. 9), membrane (not shown), bellows (Fig. 4), rolling sleeve (not shown) and the like actuators. The work of such actuators does not need further explanations.

Although above known actuators can be used, this invention provides better possibilities with the special designed member working as the above actuator or a vibrator with a constant pressurized fluid supply. In Fig. 1, such a member represents a collar hermetically fixed to the base 9 and provided with an elastic lip capable of sealing the jaw under fluid pressure in the chamber.

In an actuator work regime, the lip does not lose the sealing contact with the jaw. If the jaw ha s a larger stroke, an outlet opening is formed at the end of the stroke (at first at the lower end of the jaw where the displacement is larger). At this, the chamber is open to the surrounding atmosphere, the pressure in the chamber drops and the spring 12 pulls the jaw back. Thus, a self-oscillating regime is provided. This simplifies the construction since only a constant pressure source is necessary.

The member 8 may have a shape determined by the particular crusher, e.g. circular, right angular, square, etc. To secure a passageway on a part of the member, its thickness can be decreased in the appropriate part of its lip I, or the stiffness of the latter can be decreased. In addition, a special limiter 13, e.g. a flexible tie, can restrict the length of lip travel in a particular part (Fig. 5). Nuts 14 on the holder 15 can control the stroke of the tie, for example.

An elastic member 8 can have a flat shape and be fixed hermetically to the base 9 (Fig. 6) or the swing jaw 1 (Fig. 7). In the latter case, a limiter should be attached to the jaw since the passageway is formed between the base and the lips. The arrangement in Fig. 7 allows better ridding of foreign particles in the chamber.

The passageway can be formed also by a plurality of outlet openings 16 on the end of the lip (Fig. 8), or by orifices 17 on the lateral surface, if a diaphragm is used (Fig. 9).

II.6. CRUSHERS WITH EXPANSIONARY ACTUATORS

BACKGROUND OF THE INVENTION

This invention relates to crushers including at least two crushing members at least one of which is driven away or toward another crushing member. Such known crushers include jaw, gyratory, cone, single and multi-roll crushers. They are provided with clumsy heavy mechanical drives requiring large foundations.

The objectives of the present invention are eliminating this disadvantage, getting rid of the friction pairs in the crusher itself, simplifying its construction and decreasing the downtime.

SUMMARY OF THE INVENTION

Above objectives are achieved by the providing the driven crushing member with an actuator consisting of expansionary fluidic chambers. They are situated radially around a supporting axle and connected with a source of pressurized fluid one by one in such a manner that a progressive radial undulating movement about said axle is created progressively from one chamber to another. Said expansionary fluidic chambers represent cylinders, pipes or cavities of a resilient material, bellows, membranes, balloons, bags, etc.

The embodiments of the present invention can be made in a manner similar to known jaw, gyratory, cone and roll crushers.

IN THE DRAWINGS

Fig. 1 is the cross-sectional view of a plunger actuator of the present invention.

Fig. 2 – 4 are the same as above appropriately with radial cavities, membranes and pipes, all of a resilient material.

Fig. 5 is the face view of an actuator with bellows.

Fig. 6 is the same as above with arc diaphragms of a resilient material.

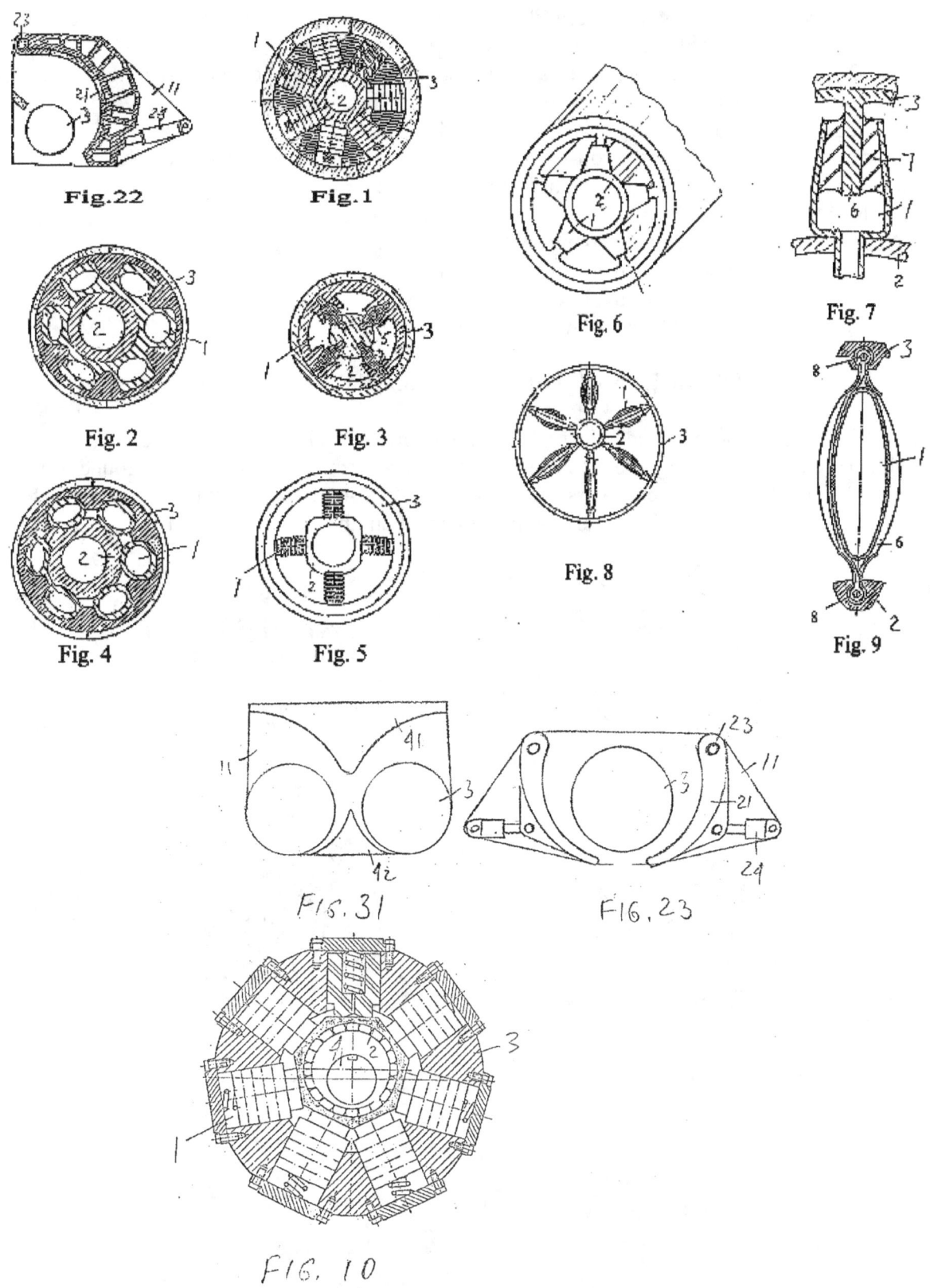

Fig.22 Fig.1 Fig. 6 Fig. 7

Fig. 2 Fig. 3 Fig. 8 Fig. 9

Fig. 4 Fig. 5

FIG. 31 FIG. 23

FIG. 10

Fig. 7 is the cross-sectional view of an arc diaphragm chamber of the above actuator.

Fig. 8 is a face view of an actuator with radial balloons.

Fig. 9 is a cross-sectional view of the above balloon.

Fig. 10 - 16 are the pulser designs corresponding accordingly to above actuators.

Fig. 17 – 19 are the cross-sectional views of single, twin and concave jaw crushers with a fluidic actuator replacing a conventional eccentric drive.

Fig. 20 is a cross-sectional view of a gyratory-like crusher with a fluidic actuator and a suspended cone.

Fig. 21 is the same as above with a cone supported with an elastic bag.

Fig. 22 and 23 are the cross-sectional views appropriately of single- and double-acting single roll-like crushers with fluidic actuators.

Fig. 24 is a diagram of a two-roll-like crusher with a fluidic actuator.

Fig. 25 is the same as above for a three-roll-like crusher with the third roll situated beneath.

Fig. 26 is the same as above with the upper third roll.

Fig. 27 is a diagram of a four-roll-like crusher with fluidic actuators.

Fig. 28 is the same as in Figure 24 with an upper stationary jaw.

Fig. 29 is the same as in Figure 25 with an upper stationary jaw.

Fig. 30 is the same as in Figure 26 with a lower stationary jaw.

Fig. 31 is the same as in Figure 28 with a lower stationary jaw.

DESCRIPTION OF THE PREFERRED EMBODIMENTS

The principal distinguishing characteristic of the present invention is that a swing crushing member is provided with an actuator consisting of expansionary fluidic chambers 1 situated radially around a supporting axle 2 (Fig. 1 - 9) in a drum 3. The chambers 1 are connected to a source of pressurized fluid (not shown) in such a manner that a progressive radial undulating movement about the axle 2 is created.

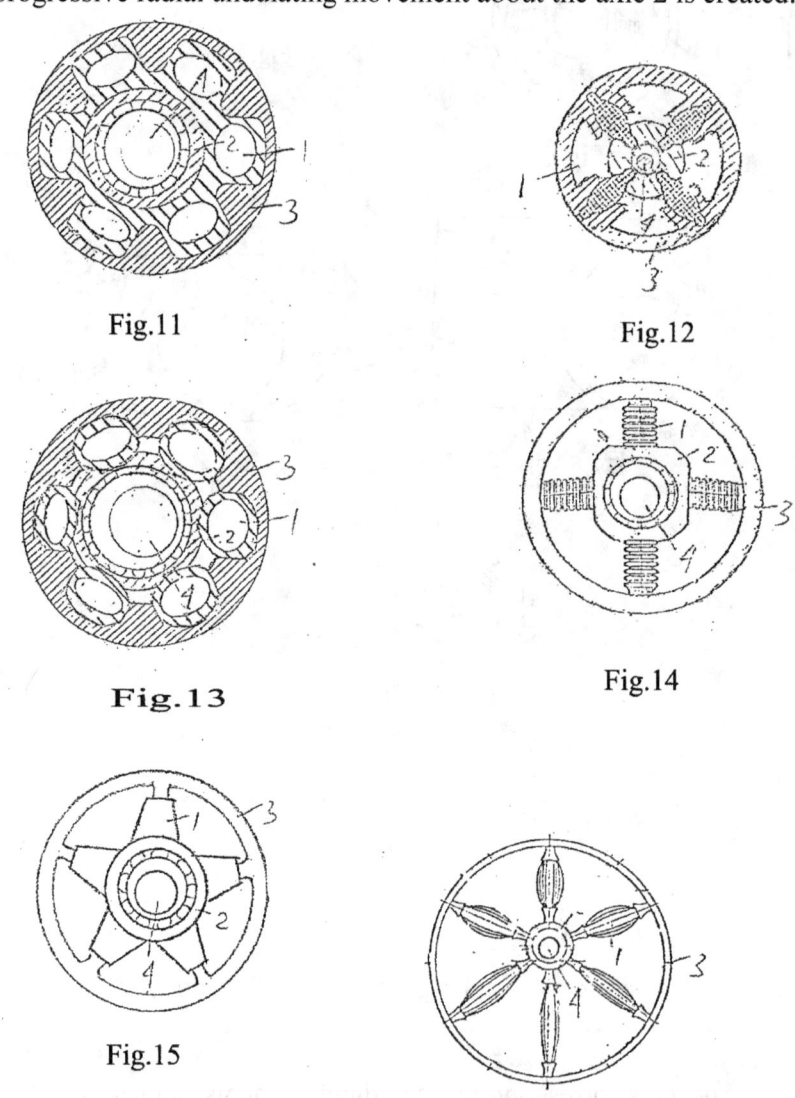

Fig.11   Fig.12

Fig.13   Fig.14

Fig.15

Fig.16

Thus, the actuator of Fig. 1 has similar features with stators of hydraulic radial-piston (motor or pump) units without built-in rotating distributor and rotor. Fluid supply of the actuator can be achieved either from a pulser with the same quantity of fluidic chambers or from a rotating distributor or valve connected to a source of a pressurized fluid.

In the first case, a pulser represents the same stator provided with an eccentrical shaft (rotor) 4 (Fig. 10 - 16) placed in a hollow axle 2. Fluidic chambers of an actuator and a pulser are interconnected accordingly with each other in such a manner that when the eccentric shaft 4 is rotating and creating radial undulating movement in the chambers of the pulser t the same movement is created in the connected chambers of the actuator.

Such a hydraulic drive is very simple and efficient. However, since the actuator and pulser are of the same order, not full economy in the mass is achieved for a relatively slow application. When the pulser shaft revolutions are lesser those of a drive (an electrical motor or internal combustion engine), the most economical means is a conventional pump with a rotating distributor or valve. In this case, the pump revolutions are of the same order of the drive and gearing, not necessarily mechanical, is provided only for the distributor or valve (an electromagnetic valve can be used).

Therefore, if to compare with conventional mechanical drives of known crushers wherein gearing is provided for the full drive, in the present case gearing is provided for the relatively small distributor which may be considered here as a fluidic amplifier. Thus, the clumsiness of the mechanical drive is eliminated.

In Fig. 2 and 3, the drum 3 is provided with cavities or pockets in which the fluidic chambers 1 are molded (Fig. 2) or represent pipes of a resilient material (Fig. 3).

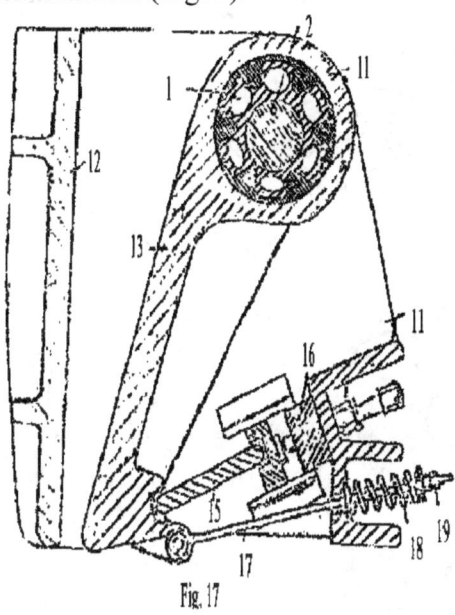

Fig. 17

In Fig. 4, the drum 3 is suspended on membranes 5 serving although not as a suspension, but as a separating member for the fluidic chambers 1. In Fig. 5, the drum 3 is suspended on bellows mounted on a polyhedron axle 2. In Fig. 6 - 7, the drum 3 is suspended on spokes 6 in rubber inserts 7, which in their turn are cast in rectangular fluidic chambers 1. In Fig. 8 - 9, the drum 3 is suspended on elastic spokes 6 with its ends fastened accordingly in the drum 3 and axle 2 by longitudinal rods 8. The spokes 6 embraces elastic balloons of the fluidic chambers 1. Here the expanding chambers stretch the spokes causing the drum to approach the axle.

The modifications of Fig. 2 - 7 have no sliding joints and therefore decrease the friction, wear and downtime. Pulsers of Fig. 10 - 16 correspond to the appropriate actuators with inserted eccentric shafts 4.

The present actuators can be built of all conventional crushers types. Non-conventional crushers can be also designed.

A jaw crusher analogous to a conventional overhead eccentric crusher is shown in Fig. 17. Therein the crusher consists of a frame 11 to which a stationary jaw 12 is fixed and a swing jaw 13 mounted on the fluidic actuator inserted in its hinge 14. The axle of the actuator is set in the frame 11. The lower end of the jaw 13 is releasable fixed by a conventional toggle arrangement and is provided with a series of shims 16. They adjust the position of the jaw 13 to vary the discharge opening between the jaws. Tension rods 17 mounted on the frame 11 hold the toggle arrangement 15 in place. They have springs 18 and threaded ends carrying nuts 19 and extended through a suitable hole in the frame11.

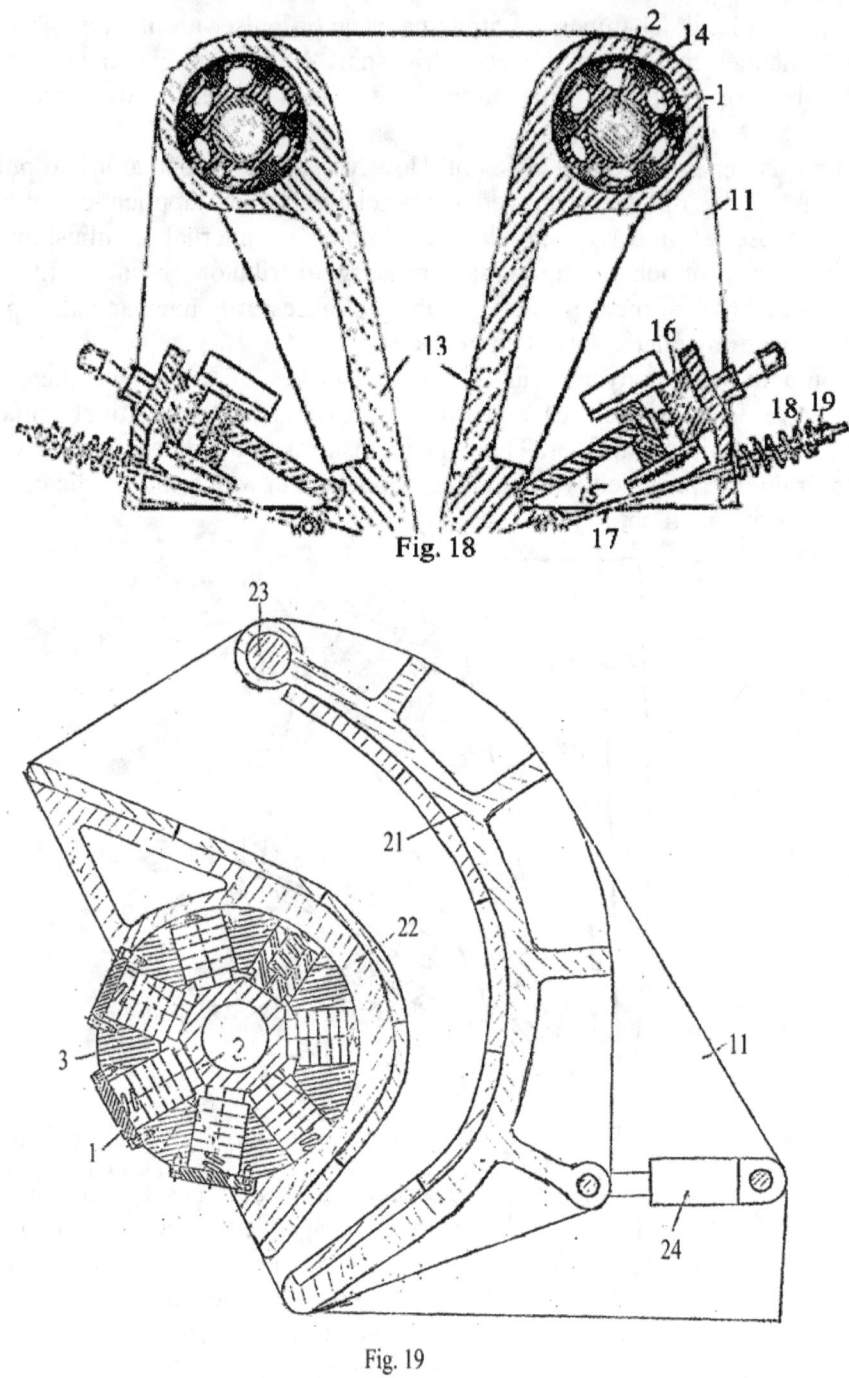

Fig. 18

Fig. 19

A jaw crusher analogous to a conventional overhead eccentric crusher is shown in Fig. 17. Therein the crusher consists of a frame 11 to which a stationary jaw 12 is fixed and a swing jaw 13 mounted on the fluidic actuator inserted in its hinge 14. The axle of the actuator is set in the frame 11. The lower end of the jaw 13 is releasable fixed by a conventional toggle arrangement and is provided with a series of shims 16. They adjust the position of the jaw 13 to vary the discharge opening between the jaws. Tension rods 17 mounted on the frame 11 hold the toggle arrangement 15 in place. They have springs 18 and threaded ends carrying nuts 19 and extended through a suitable hole in the frame11.

The rock material is crushed by the horizontal component of the actuator movement. The drawings of the crushers show one or another fluidic actuator is in way of illustration, but not in a limiting sense.

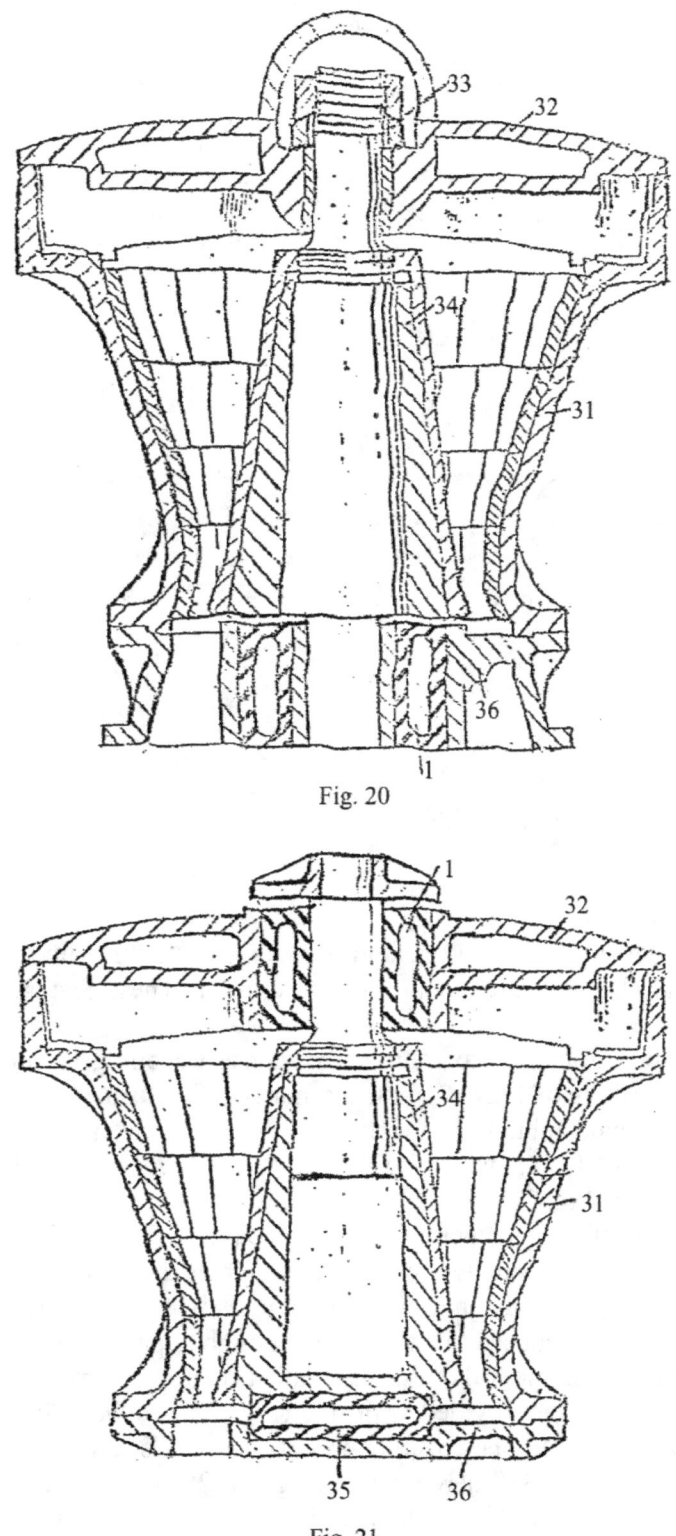

Fig. 20

Fig. 21

A twin jaw crusher (Fig. 18) consisting of two synchronized swing jaws 13 analogous to Fig. 17 does not require the further explanations.

The length of the crushing chamber is 50% - 60% more in a jaw crusher (Fig. 19) provided with a concave outer jaw 21 and a curved inner jaw 22 mounted on the fluidic actuator. Here jaws are shaped as arcs with linear extension on the inlet and outlet of the crusher.

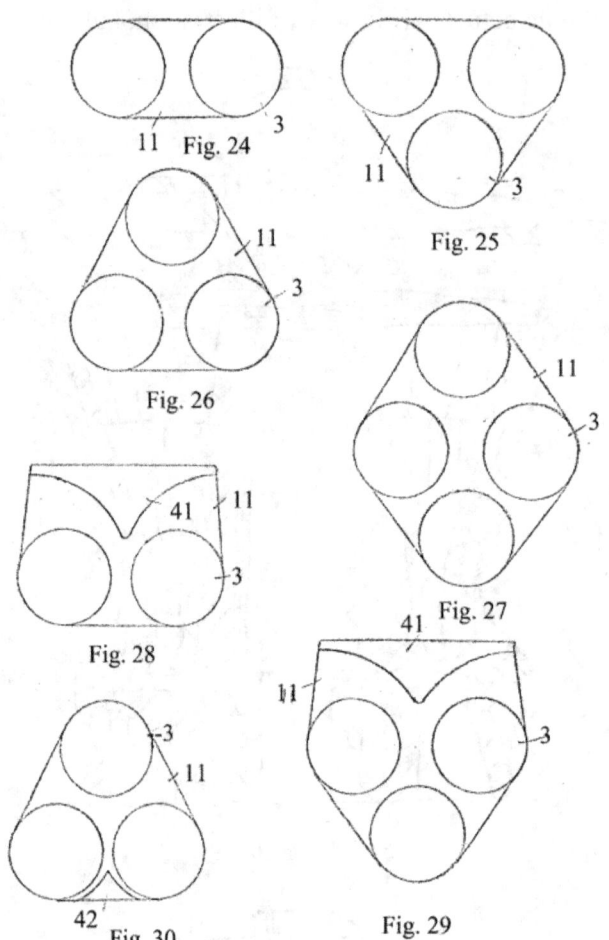

Fig. 24

Fig. 25

Fig. 26

Fig. 27

Fig. 28

Fig. 29

Fig. 30

The jaw 21 swings on a pivot joint 23 and is held in place by a hydrocylinder 24 allowing the release of non-breakable materials and changing the opening of the crusher. These features are close to conventional designs and are self-explanatory.

The actuator produces a local zone of maximum convergence travelling through the passage between the jaws. In distinction to the crushers of Fig. 17 - 18, the material is crushed here not only by the horizontal component of the actuator movement, but by the vertical component as well.

In gyratory crushers (Fig. 20 - 21), a bowl 31 is fixed with a cross-member 32 in which a spherical support 33 (Fig. 20) or a fluidic actuator (Fig. 21) for a cone 34 is inserted. In the latter case, the cone 34 is supported by an elastic bag 35 or the like fluidic chamber mounted on a base 36 fixed to the bowl 31.

The actuator swings the cone 34 about its vertical axis. When the cone approaches the bowl, the rock material is crushed and falls down under the force of gravity when the cone moves away. This process continues until the stone material is reduced to a size small enough to pass through the discharge opening.

In crushers bearing resemblance to the roll ones, the fluidic actuators are mounted instead of rolls (Fig. 22 - 31). A single roll-like crusher can be provided with one (Fig. 22) or two (Fig. 23) outer concave jaws 21 mounted in a manner similar to Fig. 19. The crushing process here is also similar.

In multi-roll-like crushers (Fig. 24 - 31) there are two (Fig. 24, 28 and 31), three (Fig. 25, 26, 29 and 30) or four (Fig. 27) synchronized actuators. The rock material fed into the clearance between the actuators is crushed, as in conventional jaw crushers, when the actuators are moving to each other, and falls down when they are going back, crushed again on the next power stroke until it finally drops out in the form of chips.

Three-actuator crushers provide twice the reduction ratio of a dual-actuator crusher. The upper additional actuator (Fig. 26, 27 and 30) provides an additional feed, the lower - an additional outlet (Fig. 25, 27 and 29). The dual feed with the upper additional actuator provides higher capacity. Both the feeds can be set differently (e.g. coarse setting and finer feed) or in the same manner. Dual outlet with the lower additional actuator has similar alternatives. The four-actuator crusher (Fig. 27) has three crushing stages and therefore provides still more capacity, the dual feed and outlet being possible.

When stationary arc V-shaped jaws 41 and 42 are installed instead of either the upper (Fig. 28, 29 and 31) or lower (Fig. 30 and 31) actuator, or both of them (Fig. 31) with their arc wings wrapped around the interacting actuators, the crushing chamber is formed in the converging passages between the arcs and actuators. The jaw increases the working area, but can wedge the material.

Although many variants are shown in the drawings, still more modifications can be made in arranging actuators and jaws within the scope of the present invention.

III IMPACT ROCK CRUSHERS

III.1. BLASTING JAW CRUSHERS

BACKGROUND OF THE INVENTION

This invention relates to jaw crushing and blasting methods of fracturing rocks. In known jaw crushers, eccentrics or vibrators drive one or both the jaws. They apply relatively static forces and transfer the energy from the drive to the jaws via massive and clumsy kinematic linkages. That requires heavy frames and foundations and prevents the creation of portable jaw crushers for underground mining.

Blasting, the cheapest method of breaking down hard rocks, cannot be used directly in most size-reduction processes. Therefore, the object of the present invention is to provide a blasting method for jaw crushing. This eliminates above disadvantages and transfers the energy directly where it is needed, without kinematic linkages. This elevates the efficiency and eliminates large masses. The proposed explosive jaw crushing method can be used in such operations where the exploitation of known crushers is economically not expedient.

Main advantages of the present method are:

great energy storage in a small volume of an energy carrier;

the possibility of using natural and artificial reservoirs;

small cost and size of the capital equipment;

easy application since blasting technique is widely used in the same field of engineering.

As to the jaws strength under the impact load, it is known (F. F. Voitsekhovskaya and B.V. Voitsekhovskiil. Fracture of Rocks by an Impact Load. Fiziko-Tekhnicheskie Problemy Razrabotki Poleznykh Iskopasmykh, Institute of Hydrodynamics, Siberian Branch of the Academy of Sciences of the USSR, Novosibirsk, No.4, July-August 1976, pp 48- 51) that the fracture energy decreases with an increase in the impact energy for a particular rock with a constant initial area of contact between the rock and its breaking tool.

The impact energy increased to several tens of thousand joules provides a sharp decrease in the tool wear, the fracture energy remaining at a low level. This is explained by the rocks heterogeneity, for example, in the change in their microhardness by a factor of 30 (granite) at points lying 1 mm apart, whereas the change in the microhardness of the tool material (steel) is less than 10% to 15%.

With low impact energies, a blunted tool cannot operate due the high fracture energy and the sharp decrease in productivity. With high impact energies and large contacting area, the fracture energy decreases by the scale effect, because there is a heightened probability of numerous strength defects existing within the zone under load. Thus, the rock breaking stress decreases. Therefore, the tooth surface of the conventional jaws is not so important in the present method since one can operate efficiently with a blunted tool. Thus, the present crushing method can be quite competitive even with the known blasting method from the point of view of the efficiency of the hard rocks breakdown.

In addition, the force on the jaw is distributed more equally herein. Therefore, the jaw is less massive and more easily subjected to dynamic processes.

SUMMARY OF THE INVENTION

The proposed method consists of placing the processed material between the jaws; providing the external surface of at least one of said jaws with a blasting charge; and detonating the latter and removing the crushed material. The charge can be placed on the jaw either directly or through a buffer medium such as rubber, air, liquid or loose friable medium. The external surface of the opposite jaw also can be provided with a similar blasting charge detonated simultaneously. Also, one or more interstitial jaws can be placed between the end jaws.

IN THE DRAWINGS

Fig. 1 - 4 are schemes of carrying out the present method with air medium, a loose friable medium, a liquid medium and with a ring charge appropriately.

Fig. 5 is a cross-sectional view of a crusher with one end burning charge.

Fig. 6 is a right side view of the crusher in Figure 5.

Fig. 7 and 8 are the same as in Fig. with two swing jaws and with an interstitial jaw.

DESCRIPTION OF THE PREFERRED EMBODIMENTS

The essence of the present method consists of a swing jaw 1 pressed against a stationary jaw 2 (Fig. 1 - 4) by a blast wave from a blasting charge 3 of a detonated explosive having a detonator 4. The wave breaks down the rocks placed between the jaws. The explosive charge can be suspended over the jaw at a predetermined (the standoff) distance).

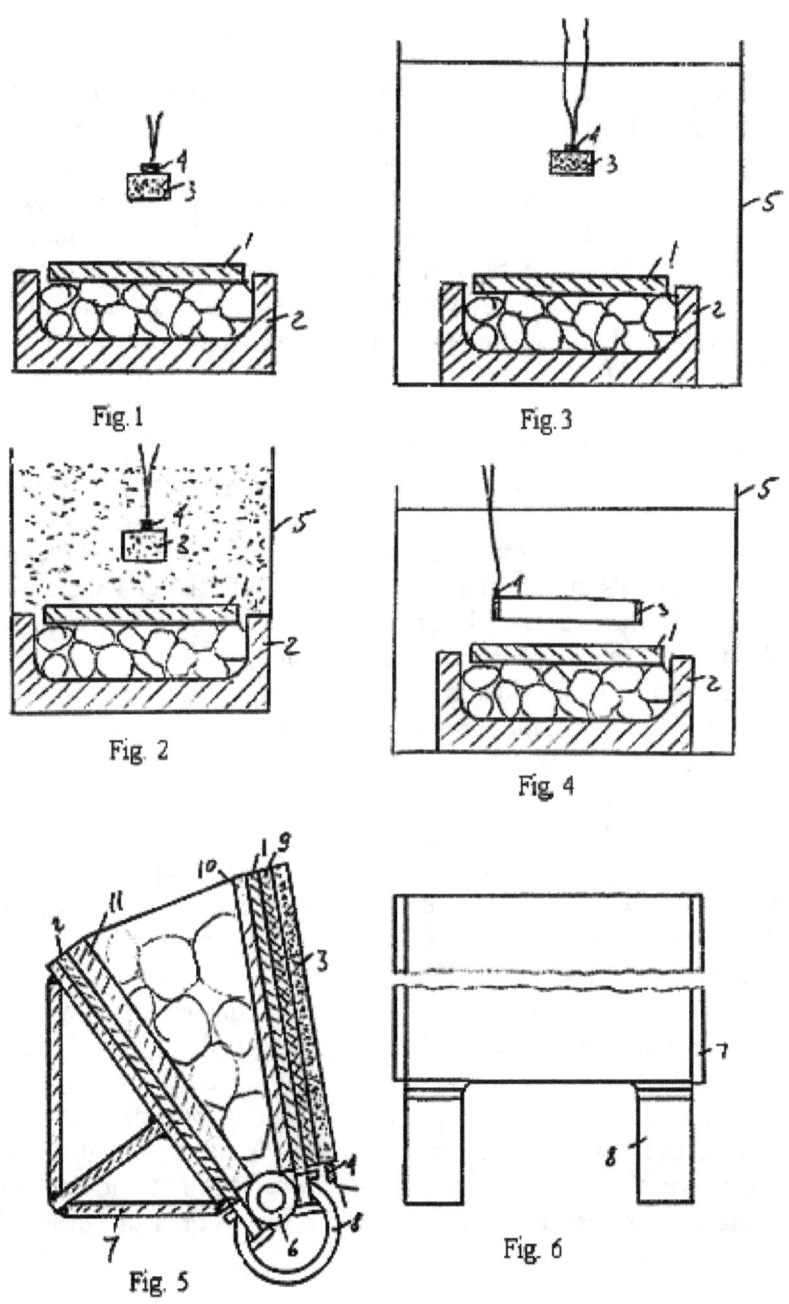

Fig. 1

Fig. 3

Fig. 2

Fig. 4

Fig. 5

Fig. 6

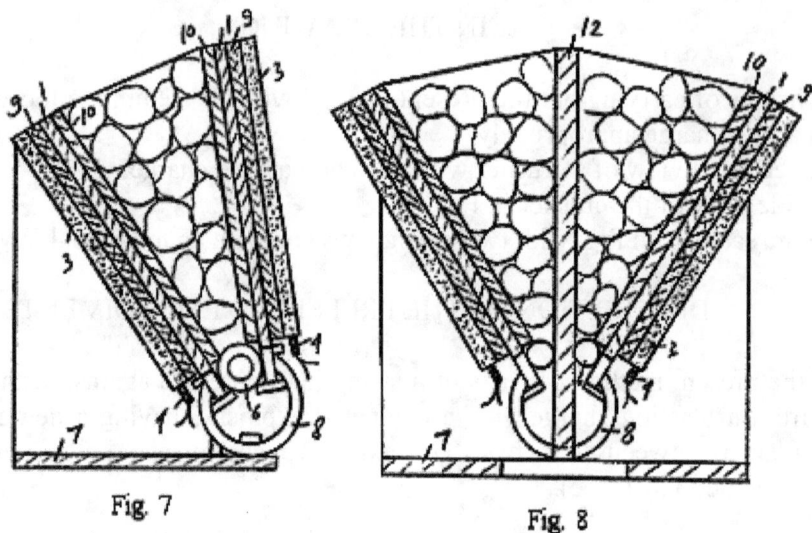

Fig. 7   Fig. 8

The complete assembly can be immersed in a tank 5 of water, or a plastic bag filled with water can be placed over the swing jaw. Also, a loose friable medium, such as crushed stones or sand can be used. A greater density of the medium increases the efficiency. That is why water is more preferable than air.

Under normal operating conditions, it is better to detonate the explosive charge as far below the surface of the water as possible, since the explosion throws the water and reduces the amount of energy lost by venting to the atmosphere of the gas bubble from the detonated charge.

The time of propagating the direct impact wave from the charge to the jaw and its acting upon the latter should be more than the time of propagating the impact wave from the charge to the water surface and of propagating the reflected wave from the latter to the jaw. This condition determines the distance between the charge and surface.

Generally, a point charge can be used for relatively small jaws. For the larger jaws, a loop of a cordlike detonating fuse consisting of a filament of an explosive material covered by a protective water-repellent coating can be placed to the outer periphery of the jaw (Fig. 4).

The water tank must be able withstanding the repeated impacts of the explosive shocks without rupturing. In large enough pools, the shocks reaching their walls are considerably reduced. To moderate the pressure of the shock wave before the walls, known methods, such as inflated rubber tubing, cushions, air bubble curtains, shock-absorbing materials, etc. can be used.

In water pool operations, a known loader is necessary to move the jaw and processed material. All the auxiliary equipment is well known in blast mining, explosive forming and the like processes.

The total impulse available for crushing is considerably less in the atmosphere than that in water, because the energy absorbed by a medium is a direct function of its density. The waves transmitted through water lose considerably more energy, than through air. The additional confinement of the explosive charge and the lengthening of the pulse due to the trapped energy, however, more than compensate for this loss of energy. Besides, several pulses can be obtained by the expansion and over-compression of the gas bubble from the explosive charge. A buffer medium (rubber, sand, etc.) distributes the pressure over the jaw more uniformly.

The movement of the jaw has two stages: a relatively small displacement with accumulating of kinetic energy and further displacement by the latter. When high explosive is detonated, the energy is liberated in two ways. Firstly, heat generates so rapidly by the chemical reaction that a shock wave is formed in the surrounding medium and this shock wave moves rapidly outwards from the detonation source. Secondly, a fluctuating gas bubble containing the products of the chemical reaction is formed and expands much more slowly than the expanding shock wave. The latter loses energy to the surrounding medium and attenuates, much less attenuation occurring in a liquid or particulate solid medium (e.g. sand, crushed rocks, etc.)

The shock wave in water represents an intense pressure wave with an extremely short rise time. Discontinuities of pressure, density, particle velocity and temperature occur across the shock front and then the high pressure decays to a normal value in very short times, usually in the order of tens of microseconds. The peak pressure in the shock wave is a function of the charge weight and falls off with the distance from the explosive. The damped exponential of the wave can decrease the pressure from 14×10^3 MN/m^2 on the charge surface t0 210 MN/m^2 on 0.15 m distance in water. Therefore, the pressure on the jaw can be in conventional limits.

The energy of the shock wave propagated with the speed of sound (1450 m/s in water) can be expressed in terms of the wave impulse and is a function of pressure and time.

The shock wave passing through the jaw loses some its energy to the jaw mass absorbing the impact as an anvil. However, since the pressure on the jaw is distributed almost uniformly (in comparison with conventional crushers where it is concentrated in a point), the mass of the jaw of the present invention is less and its damping effect can be tolerated.

The swing jaw begins moving to the opposite stationary jaw. The amount of energy imparted to the jaw depends on the peak pressure in the shock front and the time during which the shock wave takes to pass through the material, i.e. the impulse of the wave. The farther movement of the jaw proceeds under the influence of its own momentum until the crushed material absorbs the energy or until the stationary jaw or a limiter stops the swing jaw. The mass of the rocks should be taken into account as a reduced (equivalent) mass.

The impulse of the shock wave is controlled by the mass of the explosive and the standoff distance. Fluctuations of the gas bubble cause additional pressure pulses, the first pulse having 60% of the total energy, the second - 25%, the rest - 15%.

The present blasting method can be also achieved in the devices (Fig. 5 - 8) which are closer to the conventional jaw crushers with swing jaws. In a crusher with one swing jaw 1 on a cylindrical hinge 6 and fixed to the stationary jaw 2 with a frame 7 by a spring 8 (Fig. 5 and 6), the outer surface of the jaw 1 is covered with a thin layer of a one-burning-end explosive separated from the jaw by a buffer 9, e.g. rubber. The jaws 1 and 2 can have liners 10 and 11 made of high resistant steel.

The detonation wave propagating along the jaw 1 presses it to the jaw 2 with a great relative speed. To counterbalance moving masses, two opposite swing jaws can be used (Fig. 7). Here, similar cylindrical hinge 6 is fixed to a relatively light frame 7. The group charge of both swing jaws is detonated simultaneously.

In high-capacity crushers with a large angle between the jaws, an intermediate stationary jaw 12 can be placed between the swing jaws (Fig. 8). Here a mutual spring 8 fixes two cylindrical hinges 6. Such a construction represents really two single-swing jaw crushers installed oppositely, having a mutual stationary jaw and detonated simultaneously.

From the point of view of the efficiency and noise, above crushers work much better if submerged in a water pit, however, this is accompanied with technological difficulties.

III.2. COMBUSTION CRUSHERS

BACKGROUND OF THE INVENTION

This invention relates to jaw crushers. In known ones, eccentrics or vibrators drive one or two jaws, relatively static forces being applied. At the same time, at present it is known that the only reasonable inexpensive method of breaking down hard rocks is blasting which, however, cannot be used directly in most size-reduction processes.

Therefore, one objective of the present invention is to combine jaw crushers with an apparatus for creating repetitive explosions from a combustion chamber outlet by igniting hydrocarbon fuel-air mixtures. Herein a jaw represents a hinged flap or a piston hinged at the upper side of the combustion chamber and actuated from the latter. Such a design provides the direct transfer of the energy where it is needed, without kinematic linkages. This elevates the efficiency, eliminates large masses of mechanical drives and foundations and simplifies the construction. Thus, the present invention can also be related to the percussion equipment.

A question could arise as to the strength of the jaws. However, it is known (F.F. Voitsekhovskaya and B. V. Voitsekhovskii. Fracture of Rocks by an Impact Load. Fiziko-Tekhnicheskie Problemy Raztrabotki Poleznykh Iskopatmykh, Institute of Hydrodynamics, Siberian Branch, the Academy of Sciences of the USSR, Novosibirsk, No.4, July, 1976, pp. 48-51) that the fracture energy decreases with an increase in the impact energy for a particular rock with a constant initial contact area between the rock and the rock breaking tool.

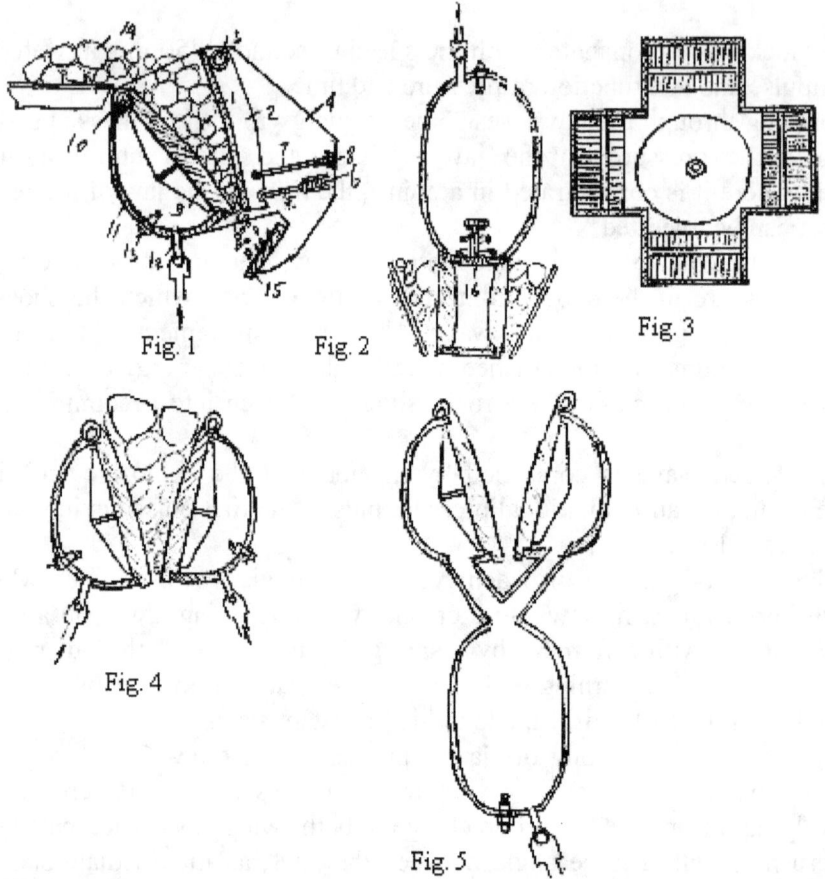

Fig. 1   Fig. 2   Fig. 3

Fig. 4   Fig. 5

The impact energy increased to several tens of thousand joules sharply decreases the wear of the tool, the fracture energy remaining at a low level. This is explained by the rocks heterogeneity, for example, in a change of their microhardness by a factor of 30 (granite) at points lying 1 mm a part, whereas the change in the microhardness of the tool material (steel) is less than 10% to 15%.

With low impact energies, a blunted tool cannot operate due to the high fracture energy and the sharp productivity decrease. With high impact energies and large area of contact, the fracture energy decreases due to the scale effect and a heightened probability of numerous strength defects existing within the loaded zone. That means that the breaking stress of the rock decreases.

Therefore, the tooth surface of the convenient jaws is not so important for the present invention, since one can operate efficiently with a blunted tool. Thus, the crusher of the present invention can be quite competitive even with the blasting method from the point of view of the efficiency of the hard rocks breaking.

IN THE DRAWINGS

Fig. 1 and 2 are sectional views of the present jaw crusher appropriately with one and two swing jaws, and one combustion chamber.

Fig. 3 is the top plan view of the present jaw crusher with four swing jaws.

Fig. 4 and 5 are the crusher of Fig. 2, provided appropriately with two separate combustion chambers, and one chamber with two combustion outlets.

THE DESCRIPTION OF THE PREFERRED EMBODIMENTS

The present jaw crusher includes a stationary jaw means 1 having a supporting frame 2 mounted for pivoted movement at its upper end via a shaft 3. The latter is conventionally supported at each end on the main frame 4. The jaw means 1 is provided with a replaceable jaw liner (not shown) of a highly wear-resistant material, such as manganese steel, and with a conventional releasable connection to the supporting frame 2.

The lower end of the stationary jaw means 1 has releasable connection with a toggle arrangement 5. The latter serves as a safety mechanism that provided with a series of shims 6. They are to adjust the position of the frame 2 pivoted about the shaft 3 in order to vary the discharge opening between the jaws.

The toggle arrangement 5 is held in place by a plurality of rods 7 pivotally mounted to the frame 2 and provided with a threaded end carrying a nut 8 and extended through a suitable hole in the main frame 4. By simple manipulation of nuts 8 and shims 6, the jaw means 1 may be pivoted and locked into a new position.

A swing jaw 9 is mounted on a pin 10 on the frame 4 and represents a hinged flap or piston for the outlet of a combustion chamber 11 charged with a combustion (preferably air-fuel) mixture from an inlet 12 and creating an explosion therein. A suitable ignition source 13 initiates combustion and creates an explosive force directed against the jaw 9.

Under the explosive blast from outlet of the combustion chamber 11, the jaw 9 moves away from the outlet and breaks down rocks 14. The exhaust blast carries the broken rocks away and impinges them against an armored plate 15 mounted on the frame 4 for the secondary breakdown and further utilizing the exhaust gases.

The combustion chamber can be provided with a valve 16 (Fig. 2) mounted before the outlet and normally held closed by the spring until a predetermined buildup of the explosive force in the chamber occurs. After that, the valve 16 quickly opens.

Fig. 2 shows the embodiment having two swing jaws to increase the efficiency. Fig. 3 illustrates a modification with four swing jaws. A counterbalanced construction with two oppositely mounted jaws swung in the opposite directions is shown in Figure 4. The same as above with a mutual combustion chamber is shown in figure 5. Therein the latter is conditionally shown in the plane of the drawing.

III.3. ELECTRIC STEAM CRUSHER

BACKGROUND OF THE INVENTION

This invention relates to rock jaw crushers. In such known crushers, eccentrics or vibrators drive one or two jaws and transfer the energy to them via massive and clumsy kinematic linkages.

The objective of the present invention is to eliminate the above disadvantage and transfer the energy directly where it is needed, without kinematic linkages. This elevates the efficiency, eliminates large masses of mechanical drives and foundations and simplifies the construction.

At present, the cheapest method of breaking down hard rocks is blasting not acceptable, however, directly in most size-reduction processes. Therefore, another objective of the present invention is to provide an impact load on the processed material, instead of relatively static forces of conventional drives.

SUMMARY OF THE INVENTION

Above objectives are achieved by combining a swing jaw with an apparatus creating repetitive steam pressure pulses or explosions from a steam chamber outlet by electrical discharges in a liquid. Such electrical treatment can be carried out by either an electrical arc between electrodes contacting the liquid, resistant heating an electroconductive liquid or electrical discharges between the droplets of non-electroconductive liquids. The swing jaw can represent a flap or piston hinged at the upper of side of said steam chamber and actuated by the latter.

As to the jaws strength under the impact load, the fracture energy decreases with an increase in the impact energy for a particular rock with a constant initial area of contact between the rock and the rock breaking tool

(F. F. Voitsekhovskaya and B. V. Voitsekhovskii, Fracture of Rocks by an Impact Load. Fiziko-Tekhnicheskie Problemy Razrabotki Poleznykh Iskopasmykh, Institute of Hydrodynamics, Siberian Branch of the Academy of Sciences of the USSR, Novosibirsk, No.4, July, August, 1976, pp 48-51). The impact energy increased to several tens of thousand joules provides a sharp decrease in the wear of the tool, the fracture energy remaining at a low level.

This is explained by the rocks heterogeneity. For example, their microhardness changes by a factor of 30 (granite) at points lying 1 mm apart, whereas the microhardness of the tool material (steel) changes in less than 10-15%.

With low impact energies, a blunted tool cannot operate owing to the high fracture energy and the sharp decrease in productivity. With the high impact energies and large area of contact, the fracture energy decreases due to the scale effect. There is a heightened probability of numerous strength defects within the zone under the load. That decreases the breaking stress of the rock. Therefore, the tooth surface strength of the convenient jaws is not so important for the present invention, since one can operate efficiently with a blunted tool. Thus, the crusher of the present invention can be quite competitive even with the blasting method from the point of view of the efficiency of the breakdown of hard rocks.

Besides, in the present invention, the force on the jaw is distributed more equally. Therefore, the jaw is less massive and more easily subjected to dynamic processes.

IN THE DRAWINGS

Fig. 1 is a sectional view of the present jaw crusher with one swing jaw.

Fig. 2 is the same as above with two swing jaws and a mutual steam chamber.

Fig. 3 is the top plan view of the present jaw crusher with four swing jaws.

Fig. 4 and 5 are the same as in Fig. 2, with two separate steam chambers and two steam outlets appropriately.

Fig. 6 is a sectional view of an electrical steam generator with a tubular electrical arc.

Fig. 7 -12 are the same as above shaped appropriately as a spark plug, with a check valve, an electromagnetic valve, a piezoelectric valve, an electrohydraulic injector and electrical spraying.

Fig. 13 is a sectional view of an electrical steam generator of a resistant heating type, with centrifugal spraying.

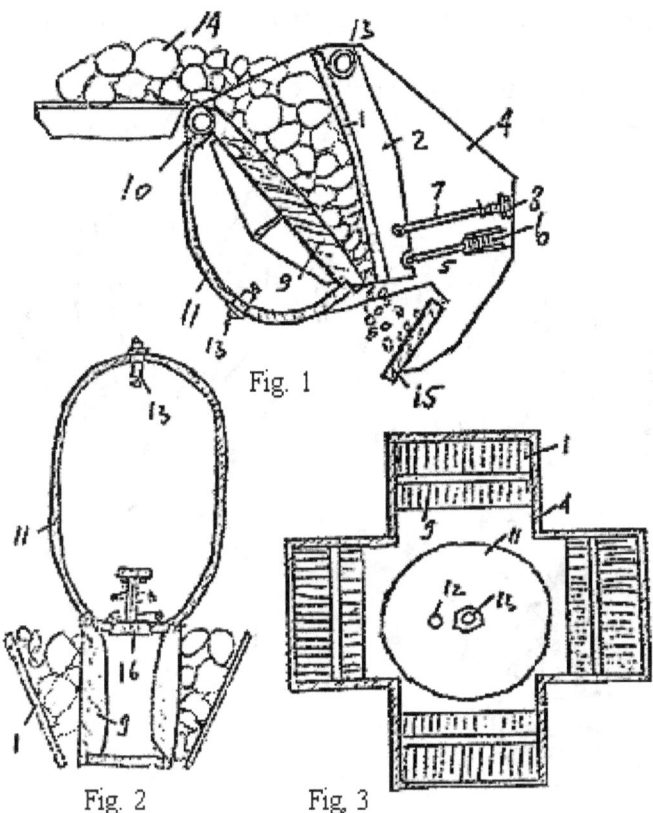

Fig. 1

Fig. 2   Fig. 3

DESCRIPTION OF THE PREFERRED EMBODIMENTS

Referring to the drawings, the jaw crusher of the present invention includes a stationary jaw means 1 including a supporting frame 2 mounted for pivoted movements at its upper end via a shaft 3. The latter, in turn, is supported at each end at the main frame 4 in a conventional manner. The jaw means 1 is provided with a replaceable jaw liner (not shown) comprising a highly wear-resistant material, such as manganese steel, for example, and releasable fixed to the supporting frame 2 by usual means.

The lower end of the stationary jaw means 1 is releasable fixed in position by a toggle arrangement 5 serving as a safety mechanism and provided with series of shims 6 to adjust the position of the frame 2 pivoted about the shaft 3 to vary the discharge opening between the jaws. The toggle arrangement 5 is held in place by a plurality of rods 7 pivotally mounted to the frame 2 and provided with a threaded end carrying a nut 8 and extended through a suitable hole in the main frame 4. By simple manipulation of nuts 8 and shims 6, the jaw means 1 may be pivoted and locked into a new position. A swing jaw 9 is mounted on a pin 10 on the frame 4 and represents a hinged flap or piston for the outlet of a steam chamber 11 having a suitable electrical steam generator 13 creating an explosive, dynamic or relatively static force directly against the jaw 9.

When steam pressure is applied to the back of the jaw 9, the latter moves out of the outlet of the chamber 11 and breaks down rocks 14, the steam expanding out of the outlet. At this, the exhaust steam carries broken down rocks away and impinges them against an armored plate 15 mounted on the frame 4 for the secondary breakdown and further utilizing of the exhaust steam. The steam chamber can be provided with a valve 16 (Fig. 2) mounted before the outlet and normally held closed by the spring until a predetermined build-up in the explosive force in the chamber occurs, after which the valve 16 quickly opens.

Fig. 2 also illustrates a modification of the present invention with two swing jaws to increase the efficiency. Fig. 3 illustrates a modification with four swing jaws. A counterbalanced design with two opposite jaws swung in the opposite directions is shown in Fig. 4.

Fig. 5 illustrates the same as above with a mutual steam chamber. Therein the latter is conditionally shown in the plane of the drawing.

The electrical steam generator 13 can be executed in many variants. In Fig. 6, a pipe 21 made of an insulator is mounted inside a hollow electrode 22 connected to a source of electrical current, to which a disc electrode

23 is also connected via a leg 24 and a wall 25 of the chamber 11. The electrode 22 is insulated from the wall 25 by an insulator collar 26.

A working liquid (e.g. water) passing through the zone of arc burning is heated by the surrounding tubular arc between the electrodes 22 and 23. Overheated liquid jet is exploded with vapor.

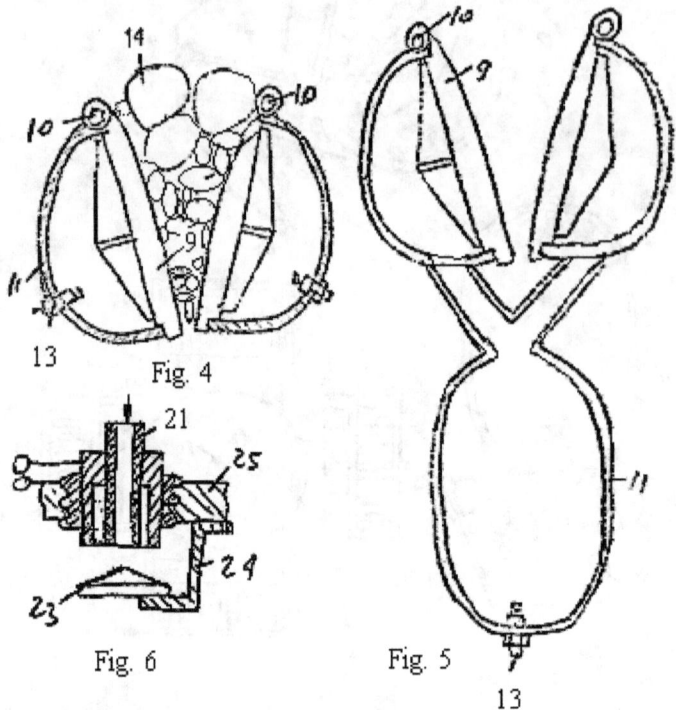

Fig. 4

Fig. 6

Fig. 5

13

The steam generator 13 can also have the shape of a spark plug (Fig. 7) consisting of a body 27 with a central electrode 28 and a cap 29 provided with a pipe 30 for a working liquid. The cap 29 and pipe 30 represent also a mass electrode. The body 27 and electrode 28 are insulated with a collar 31. The electrodes are connected to a high voltage system (not shown).

The liquid passes via the pipe 30 and spreads on the cap 29 as a film. A spark discharge between the electrode 28 and cap 29 vaporizes the liquid.

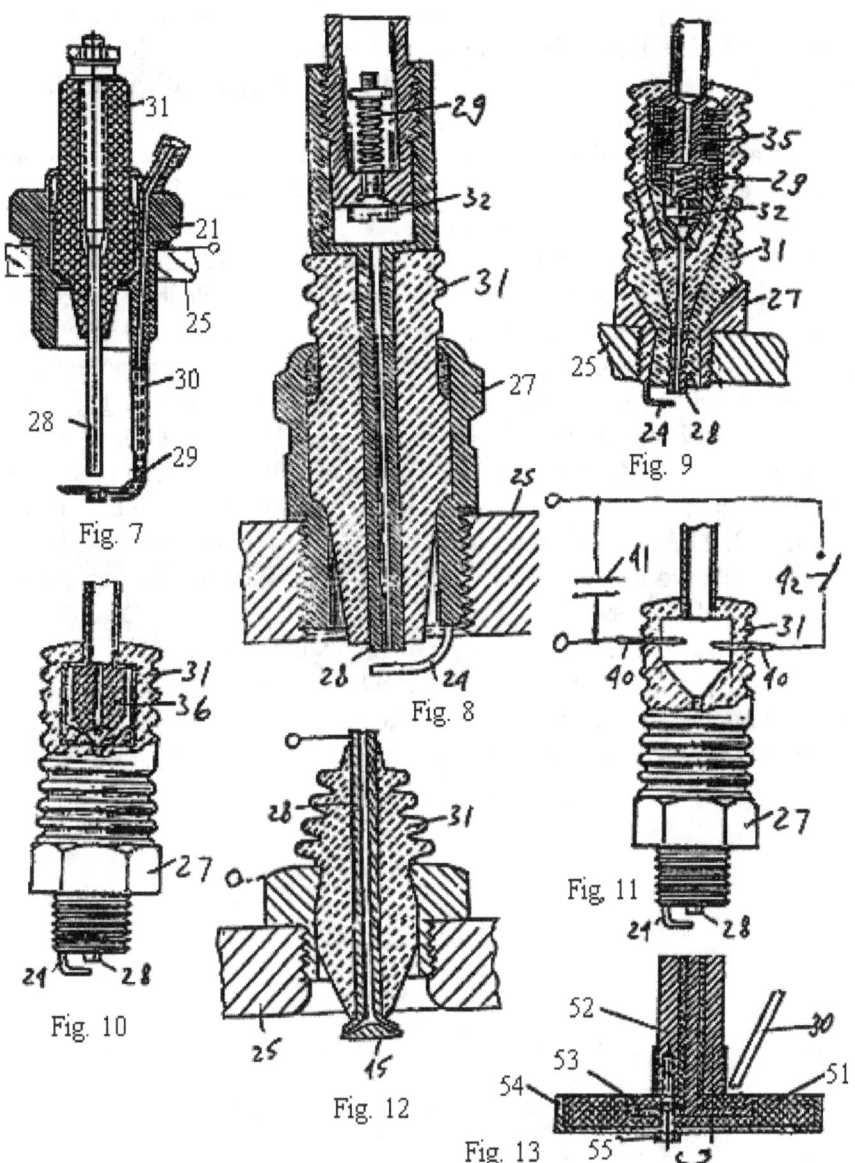

Fig. 7

Fig. 8

Fig. 9

Fig. 10

Fig. 11

Fig. 12

Fig. 13

The generator in Fig. 8 has a similar construction, the difference lying in a hollow central electrode 28 with a channel for a liquid, a leg electrode 34 and a check spring-loaded valve 32 in the inlet of the said channel. The valve is open under the pressure of a liquid-feed system (not shown) and closed under the spring action when the steam pressure is higher than the first pressure. Thus, the valve secures the pulsing feeding.

The generators of Fig. 9 and 10 have similar features, the difference lying in an electromagnetic valve (Fig. 9) and piezoelectric-transducer valve (Fig. 10), the electromagnet 35 and transducer 36. The latter represents a valve body of piezoelectric ceramics connected in parallel with the arc ignition system.

The energized electromagnet 35 (Fig. 9) draws the valve 32 upward against the normal bias of the spring 29 opening the valve and permitting the liquid into an electrical discharge gap wherein the liquid is evaporated under the ignited arc. In Fig. 10, the transducer valve 32 contracts in length in response to the transmission of a current there through and lets the liquid to pass into the discharge gap.

Still another means for pressing the liquid into the discharge gap is electrohydraulic impacts of electrical discharges between electrodes 40 placed in a cavity of the collar 31 (Fig. 11). A typical electric power unit contains a capacitor 41 connected in series with a switch 42. When the latter is open, the capacitor 41 stores some electrical energy from a power source. Then the switch 42 is closed, the stored electrical energy is discharged across the electrodes 40, produces a hydraulic shock and presses the liquid into the electrical arc gap between the electrode 28 and leg 24.

For a better evaporation, the liquid can be discharged as small droplets produced by a corona discharge.

For this, a direct high voltage is maintained between the wall 25 and a hollow electrode 28 provided with a spraying plate 45 (Fig. 12). The liquid passing the channel of the electrode 28 picks up some charge from it. The charge causes repulsion of the liquid from the channel through the gap formed by the plate 45 and the droplets. The internal repulsive forces of the charged individual droplets disperse them farther into smaller and smaller droplets evaporating faster.

The evaporation can be achieved by the ignition of an electrical arc as above. The sudden reverse of the polarity of the electrode 28 and the mass causes a sudden electron current through the liquid mist, creating an ignition and resulting in almost the explosive evaporation.

In all above cases, an electrical arc between metallic electrodes is used for instant heating of the liquid (in the last case more complicated phenomenon occurs. However, if an electroconductive liquid is used, the liquid jet can serve as an electrode, the second metallic electrode being not necessary. All above steam generators can be used for such a work without modifications. The author used an electrolytic electrode in his plasma torch (S.I. Fishgal. Plasma Arc-Cutting of Metals. Soviet Invention No. 195,299; S.I. Fishgal. Liquid Jet as an Electrode. Inventor and Rationalizer. Moscow, 1967, No. 8. These works are in Addendum.).

Of course, it is not necessary to produce an arc, since an electrical energy can be converted into heat by the resistance of the liquid conductor. An example of such a construction using also centrifugal forces for spraying is shown in Fig. 13. Therein a spraying disc 51 of an insulating material is mounted on a rotating shaft 52 and has cap-shaped electrodes 53 and 54 connected to an electrical power source by sliding contacts (not shown). Screws 55 fasten all the assembly together.

An electroconductive liquid is passed from the pipe 30 on the rotating disc 51 and forms a film. The electrical current passing through the latter between the electrodes 53 and 54 heats the liquid up to boiling and evaporation. The evaporation rate depends on the voltage applied.

Many other variants and modifications are possible within the scope and spirit of the present invention.

IV ADDENDUM

IV.1. SPRING IN PECULIAR CASES
Inventor and Rationalizer/Innovator. Moscow, 1968, No. 3
(n Russian)

ABSTRACT

The author used springs to form cavities in concrete structures for easier extracting by winding up one spring end. In a friction welding, a helical spring transformed linear motions into the torsion ones. That spring in a hose served as a flexible screw conveyer, and two of them placed concentrically worked as a mixer too. A helical spring welded on a cylinder was used as a screw gear transmission. A belt pulley in transmissions and conveyors with a welded spring wound in the opposite directions prevented the belt skidding. Bending pipes is easier with a spring insert.

КОНСТРУКТОРСКИЕ НАХОДКИ

С. ФИШГАЛ,
инженер.

Пружина в особых случаях

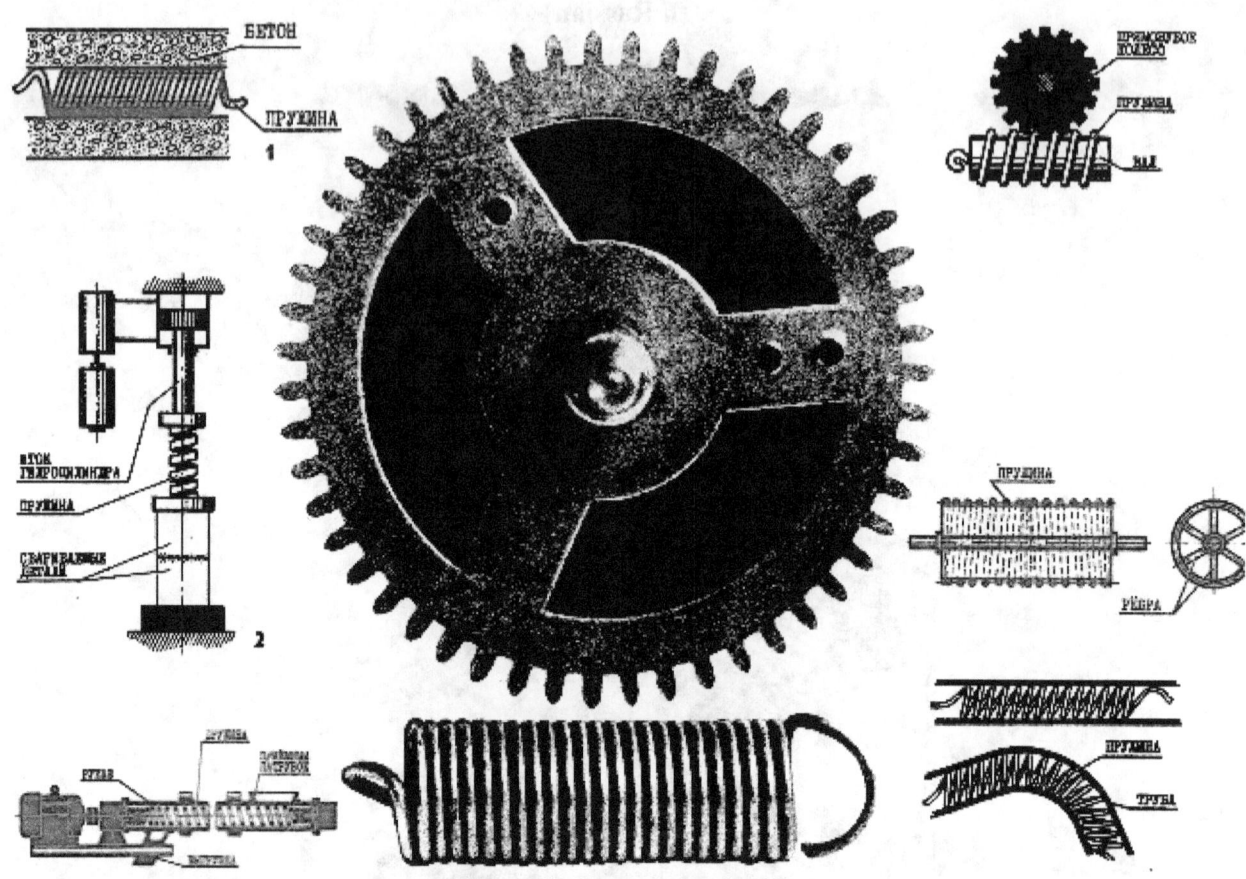

Свое основное назначение — аккумулировать механическую энергию — пружина приобрела, когда человек впервые согнул ветку для капкана. Появились кареты и кровати — пружина стала амортизатором. Впоследствии — измерительным прибором. Другие возможности пружин известны менее широко.

О двух особых назначениях пружины — для упрочения обечаек и в качестве винтового двигателя — журнал сообщал (ИР, 2, 67). Приведем еще несколько примеров использования винтовых пружин в различных конструкциях.

Пустоты в железобетонных изделиях обычно делают с помощью полых металлических стержней. Но стержни схватываются с бетоном, и тогда их трудно вытащить после формовки, не повредив самого бетона. Чехословацкие инженеры заменили сплошные цилиндры спиральной пружиной (рис. 1). Ее навивают очень плотно, виток к витку, и заливают бетоном. Пружина тоже срастается с готовой деталью, но извлечь ее значительно легче. Для этого один из концов пружины вращают, как бы для скру-

чивания, и витки поочередно уменьшают свой диаметр. Спираль высвобождается из бетона с гораздо меньшим усилием.

Для вибрационного метода сварки потребовался механизм, способный совмещать возвратно-поступательное «дрожание» с давлением на обрабатываемую поверхность. Эту задачу успешно решила винтовая пружина (рис. 2). Это делается так. В гидравлическом цилиндре создают пульсирующее давление. Через поршень оно передается на подпружиненный шток. Во время такта сжатия пружина стремится раскрутиться, при растяжении — закрутиться. Таким образом пружина давит на деталь, одновременно сообщая ей крутительные колебания.

Спиральную пружину можно использовать в виде гибкого шнека (рис. 3). В этом случае ее помещают в рукав, на котором монтируют приемный и выгрузной патрубки. Такой шнек будет с успехом выгружать сыпучие материалы, например, из крытых вагонов. А если в гибкий рукав вставить еще одну спираль, то транспортер будет еще и смесителем.

Червячные передачи, как правило, сложны в изготовлении. Сделать червяк можно, навивая на стержень проволоку без зазора, виток к витку (рис. 4). А прямозубую шестерню вместо колеса отштампуйте из тонкой листовой заготовки.

Конвейерные ленты и плоские ремни имеют неприятное свойство соскакивать с барабанов и шкивов. Нередко их ограждают различными устройствами, не зная о существовании простого и надежного предохранителя. Намотайте на стержень или пустотелый барабан проволоку, но так, чтобы витки имели наклон в разные стороны (рис. 5). С такого барабана конвейерная лента никогда не соскочит.

Хотелось бы подсказать еще одну возможность массового применения пружин. При гибке труб их предварительно заполняют песком, который приходится выбивать или выдувать. Вставьте в изгибаемую трубку пружину (рис. 6) и вы избавитесь от этих хлопот. Да и качество колена на трубе намного повысится.

г. Киев

IV.2. HYDRAULIC SHOCK CALCULATIONS

(S.I. Fishgal, The Calculation of Pistonless Hydroshock Devices, Soviet Applied Mechanics, Consultants Bureau, New York, June, 1975)

Russian Original Vol. 9, No. 11, November, 1973

June 1, 1975

SOAMBT 9(11) 1153-1270 (1973)

SOVIET
APPLIED MECHANICS

ПРИКЛАДНАЯ МЕХАНИКА
(PRIKLADNAYA MEKHANIKA)

TRANSLATED FROM RUSSIAN

 CONSULTANTS BUREAU, NEW YORK

ABSTRACT

The calculation of pistonless hydraulic shock devices that may include a hydraulic actuator, a pump, a distributor, an accumulator, a multiplier, a pipe system, and so on. They are used in nut-runners, vibrators, testing benches, the treatment of tubes, etc.

THE CALCULATION OF PISTONLESS HYDROSHOCK DEVICES

S. I. Fishgal   UDC 621.22.011

Hydroshock devices with a slave mechanism in the form of a system of pipelines (calculating schemes are shown on Fig. 1) are used in the hydrodynamic treatment of tubes, in wrenches, vibrators, industrial apparatus, research units, etc. The sources of pressure in such units are vibropumps with a separate or combined distributor, hydroaccumulators with a rapid-acting distributor (devices of type a on Fig. 1), or hydromultipliers with a pneumatic-hammer or similar shock mechanism (devices of type b on Fig. 1).

With a pulsating pressure, there arise transverse vibrations of the tube, which are desirable for vibrofeeders, but which are not admissible in the hydroshock treatment of tubes, due to the possibility of their fatigue failure. The amplitude of the pressure pulsation, when there is resonance with the column of liquid in the pipeline, increases considerably. This resonance frequency can be determined using the formulas of [2].

Investigations have shown the insignificant effect of fluctuations of the velocity of the liquid, within the range of velocities employed in machine building, on the dynamic stability of pipelines, in comparison with pulsations of the pressure.

In calculation of a hydroshock device, according to the purpose for which it is intended, maximal values are assigned to the change in the radius ΔR_M of the tube and the frequency f of the pulsations.

The maximal pressure, capacity, and power are determined using the formulas

$$P_{M} = \frac{2\Delta R_{M} E \delta}{R_{0}^{2}(1 - 0.5\mu_{p})}; \quad Q_{M} = \pi \Delta R_{M} \frac{L}{t}(2R_{0} + \Delta R_{M}); \quad N_{M} = Q_{M} P_{M},$$

where R_0, δ, E, μ_p, L are the initial inside radius, the elastic modulus, the Poisson coefficient, and the length of the pipeline; t is the time of action of the pressure on the pipeline.

In the case of the use of tubular springs in vibrators with torsional vibrations, the angle of rotation of the end of the spring under the action of the pressure, the central angle of the helical spring, and the change in volume of the internal cavity ΔV (the mass flow rate of liquid for one pulse) are determined from [1].

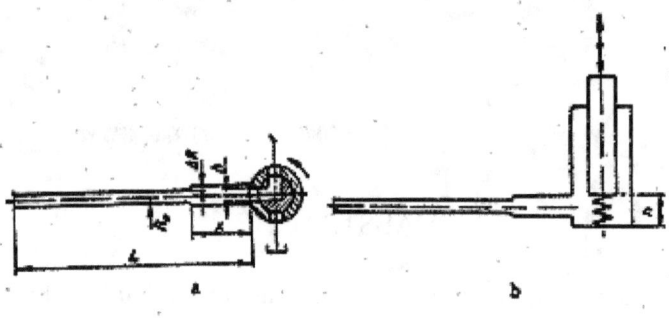

Fig. 1

Stroidormash Plant, Kiev. Translated from Prikladnaya Mekhanika, Vol. 9, No. 11, pp. 125–129, November, 1973. Original article submitted April 17, 1972.

© 1975 Plenum Publishing Corporation, 227 West 17th Street, New York, N.Y. 10011. No part of this publication may be reproduced, stored in a retrieval system, or transmitted, in any form or by any means, electronic, mechanical, photocopying, microfilming, recording or otherwise, without written permission of the publisher. A copy of this article is available from the publisher for $15.00.

The work of the source of pressure for one pulse is equal to

$$A = p\Delta V.$$

In a dynamic calculation of a hydroshock device, the structural parameters and the working characteristics are made more precise, and the stability of its operation is verified. The compressibility of the liquid is not taken into account with $p > 2$ MN/cm^2 [6]; wave processes are not taken into consideration with a length of the pipelines which is not commensurate with the wavelength, or with a frequency of the vibrations which is not commensurate with the period of the pipeline (in practice, with $f < 50$ Hz).

The equation of the expansion of a dead-end pipeline under the action of pressure has the form

$$m(\Delta R)'' + \frac{4\pi E\delta L\Delta R}{R_0(1 - 0.5\pi_p)} + P - pF = 0 \tag{1}$$

Here m is the mass of the pipeline and the movable elements of the load; P is the constant component of the working load.

On the basis of the wave equations for a hydroshock [4], we obtain

$$p = p_0 + \rho a\,[\varphi(x - at) + \psi(x + at)];$$

$$v - v_0 = \varphi(x - at) - \psi(x + at).$$

where x is a coordinate along the length of the pipeline; v and v_0 are the instantaneous and initial velocities of the liquid in the pipeline; $\varphi(x - at)$ and $\psi(x + at)$ are arbitrary functions of the forward and back waves, depending on the boundary conditions; ρ is the density of the liquid; a is the velocity of the shock wave.

The forward wave, forming at the start of the pipeline $(x = 0)$, propagates along it with the velocity a. After the time $t = L/a$, it reaches the end of the pipeline $(x = L)$, as a result of which there is formed a reflected back wave, propagating at the same velocity in the opposite direction. Thus, at any given moment of time, the values of the functions of the forward and back waves are equal, respectively, to the values of the functions of these waves in the bounding cross sections of the pipeline

$$\varphi(x - at) = \varphi\left[-a\left(t - \frac{x}{a}\right)\right];$$

$$\psi(x + at) = \psi\left[L + a\left(t - \frac{L - x}{a}\right)\right].$$

Let us consider the instantaneous pressure p_i and velocity v_i after the period of time $t = L/a$

$$p_i = p_0 + \rho a(\varphi_i + \psi_{i-1}); \qquad v_i - v_0 = \varphi_i - \psi_{i-1} \quad (x = 0);$$

$$p_i = p_0(\varphi_{i-1} + \psi_i); \qquad v_i - v_0 = \varphi_{i-1} - \psi_i \quad (x = L). \tag{2}$$

The equation of the mass flow rate of hydroshock devices with $x = L = 0$, when the length of the pipeline is not great and there are no wave processes, has the form

type a

$$2\pi R_0 L(\Delta R)'\dot{p}\frac{V_L}{E_L} - \mu b y\sqrt{2\frac{p}{\rho}} = 0; \tag{3}$$

type b

$$2\pi R_0 L(\Delta R)' - \dot{p}\frac{V_L}{E_L} - Fw = 0. \tag{4}$$

Here E_L is the elastic modulus of the working liquid; μ, b, y are the coefficient of the mass flow rate, and the width and instantaneous height of the working window of the distributor; $y = h\sin\omega_d t = h\omega_d t$; b is the height of the working window of the distributor; ω_d is the angular velocity of the distributor (the frequency of the vibrations of the valve-type pulsator); V_L is the volume of liquid in the pipeline (since the deformation of the latter is not great, we assume $V_0 = $ const); F, w are the area and the acceleration of the plunger of the multiplicator, on which the shock mechanism acts.

If $E_L = \infty$ (the liquid is incompressible) and there are no wave processes, we then obtain:

type a

$$m(\Delta R)^{\cdot\cdot} - [(\Delta R)^{\cdot}]^2 a_1 \operatorname{cosec}^2 \omega_d t + Z\Delta R = -P; \tag{5}$$

type b

$$\Delta R = 0.08 \omega t^2 F R_0^{-1} L^{-1}; \tag{6}$$

$$p = \omega F R_0^{-2}\left(0.025 m L^{-2} + \frac{E\delta t^2}{1 - 0.5\mu_p}\right) + 0.64 P R_0^{-1} L^{-1}. \tag{7}$$

Here

$$a_1 = 122 R_0^{\frac{1}{3}} L^3 \rho \, (\mu b h)^{-2}; \quad Z = \frac{12.5 E\delta L}{R_0(1 - 0.5\mu_p)}.$$

To investigate the stability of the operation, we linearize Eq. (5) by the method of piecewise-linear approximation [5]

$$m(\Delta R)^{\cdot\cdot} - A_1 \Delta R_M a_1 \operatorname{cosec}^2 \omega_p t T^{-1} (\Delta R)^{\cdot} + Z\Delta R = -P - B\Delta R_M^2 T^{-2} a_1 \operatorname{cosec}^2 \omega_p t, \tag{5}$$

where

$$A_1 = 0.38; \quad B = 0 \quad \text{with } 0 \leqslant (\Delta R)^{\cdot} \leqslant 0.46\frac{\Delta R_M}{T};$$

$$A_1 = 1.46; \quad B = 0.5 \quad \text{with } 0.46\frac{\Delta R_M}{T} \leqslant (\Delta R)^{\cdot} \leqslant \frac{\Delta R_M}{T};$$

T is the time of the total displacement of the wall of the pipeline.

For the work of the system under vibrational conditions, in accordance with the Gurvich criterion, the coefficients of the equation must be negative, which is ensured in the given case.

Taking account of the compressibility of the liquid, but not taking account of wave phenomena, using the method of piecewise-linear approximation [5] we obtain:

type a

$$m(\Delta R)^{\cdots} - A_2 a_2 m \sin\omega_d t (\Delta R)^{\cdot\cdot} + \left(39.5 R_0^2 L^2 \frac{E_L}{V_L} + Z\right)(\Delta R)^{\cdot}$$
$$- A_2 a_2 Z \sin\omega_d t \Delta R = A_2 a_2 \sin\omega_d t (P + 6.28 R_0 L B_1 p_M); \tag{9}$$

type b

$$m(\Delta R)^{\cdots} + \left(39.5 R_0 L^2 \frac{E_L}{V_L} + Z\right)(\Delta R)^{\cdot} = 6.28 R_0 L F E_L V_L^{-1} \omega t. \tag{10}$$

Here

$$A_2 = 4.53; \quad B_1 = 0 \quad \text{with } 0 \leqslant p \leqslant 0.07 p_M;$$

$$A_2 = 0.78; \quad B_1 = 0.27 \quad \text{with } 0.07 p_M \leqslant p \leqslant p_M;$$

$$a_2 = \frac{1.41 E_L \mu b h}{V_L \sqrt{\rho p_M}}.$$

An analysis of the equations obtained shows that a vibrational process is ensured by the construction of the devices. It must be noted that the error of the approximation does not exceed 4% of the maximal value of the approximated function.

Taking account of wave processes and neglecting the compressibility of the liquid, we determine the boundary conditions for the start of the pipeline:

type a

$$\mu b y \sqrt{2\frac{p_t}{\rho}} + \pi R_0^2 v_t = 0; \tag{11}$$

type b

$$F\omega t + \pi R_0^2 (v_t - v_0) = 0. \tag{12}$$

Simultaneous solution with Eqs. (2) gives:

type a

$$\pi R_0^2 \left(2\psi_{i-1} - \frac{p_i - p_0}{\rho a}\right) - \mu b y \sqrt{2 \frac{p_i}{\rho}} = 0; \tag{13}$$

type b

$$\pi R_0^2 \left(2\psi_{i-1} - \frac{p_i - p_0}{\rho a}\right) - F w l = 0. \tag{14}$$

Knowing the function of the backward wave, using these equations we can determine the increase in the pressure at the start of the pipeline for any given period (for example, by the graphic method of [3]).

The function of the forward wave for the cross sections $x = 0$ and $x = L$ is equal to

$$\varphi_i = \begin{vmatrix} \frac{p_i - p_0}{\rho a} - \psi_{i-1} & (x = 0); \\ \varphi_{i-1} - (v_i - v_0) & (x = L). \end{vmatrix} \tag{15}$$

Solving Eqs. (13), (14), (15) simultaneously, taking account of the initial conditions (in the first period, there is no backward wave), we can obtain the change in the pressure in the pipeline after an interval of time. The displacement of the wall of the pipeline as a function of time is determined by substitution of the values found into Eq. (1).

With $p < 2$ MN/m^2, wave processes are rapidly damped, and are not taken into consideration.

LITERATURE CITED

1. L. E. Andreeva, Elastic Elements of Instruments [in Russian], Izd. Mashgiz, Moscow (1962).
2. T. M. Bashta, Machine-Building Hydraulics [in Russian], Izd. Mashinostroenie, Moscow (1971).
3. A. E. Zhmud', Hydraulic Shock in Hydroturbine Units [in Russian], Izd. Gosénergoizdat, Moscow —Leningrad (1953).
4. N. E. Zhukovskii, Hydraulic Shock in Water Pipelines [in Russian], Izd. Gostékhizdat, Moscow —Leningrad (1949).
5. S. I. Fishgal, "The approximation of functions of the resistance and the mass flow rate," Prikl. Mekhan., 7, No. 9 (1971).
6. S. I. Fishgal, "Hydroshock determination of the vapor—gas content of a liquid," Zavod. Lab., No. 12 (1972).

IV.3. HYDRAULIC IMPACT COMPUTATION
(S.I. Fishgal. Re Computation of Pistonless Hydroimpact Devices. Applied Mechanics, v. 9, Kiev, 1973, No. 11)

NOTE. Consultant Bureau of New York translated from Russian and published this paper in S.I. Fishgal, The Calculation of Pistonless Hydroshock Devices, Soviet Applied Mechanics, Consultants Bureau, New York, June, 1975.

ABSTRACT

The calculation of pistonless hydraulic impact (shock) devices is given. Such devices may include a hydraulic actuator, a pump, a distributor, an accumulator, a multiplier, a pipe system, and so on. They are used in the treatment of tubes, nut-runners, vibrators, testing benches, etc.

УДК 621.22.011

К РАСЧЕТУ БЕСПОРШНЕВЫХ ГИДРОУДАРНЫХ УСТРОЙСТВ

С. И. Фишгал

(Киев)

Гидроударные устройства с исполнительным механизмом в виде системы трубопроводов (расчетные схемы показаны на рисунке) применяются при гидродинамической обработке труб, в гайковертах, вибраторах, технологических аппаратах, испытательных установках и др. Источником давления в таких устройствах являются вибронасос с отдельным или совмещенным с ним распределителем, гидроаккумулятор с быстродействующим распределителем (устройства типа *а* на рисунке) и гидромультипликатор с пневмомолотком или подобным ему ударным механизмом (устройства типа *б* на рисунке).

При пульсирующем давлении возникают поперечные колебания трубопровода, которые желательны для виброприводов и не допустимы при гидроударной обработке труб из-за возможности их усталостного разрушения. Амплитуда колебаний давления при резонансе со столбом жидкости в трубопроводе значительно увеличивается. Эту резонансную частоту можно определить по формулам работы [2].

Исследования показали незначительное влияние колебаний скорости жидкости в пределах применяемого в машиностроении диапазона скоростей на динамическую устойчивость трубопроводов по сравнению с пульсацией давления.

При расчете гидроударного устройства в соответствии с его целевым назначением задают максимальные значения изменения радиуса $\Delta R_{\text{м}}$ трубы и частоту f колебаний.

Максимальное давление, производительность и мощность определяются по формулам

$$p_{\text{м}} = \frac{2\Delta R_{\text{м}} E \delta}{R_0^2 (1 - 0,5\mu_{\text{п}})}; \quad Q_{\text{м}} = \pi \Delta R_{\text{м}} \frac{L}{t} (2R_0 + \Delta R_{\text{м}}); \quad N_{\text{м}} = Q_{\text{м}} p_{\text{м}},$$

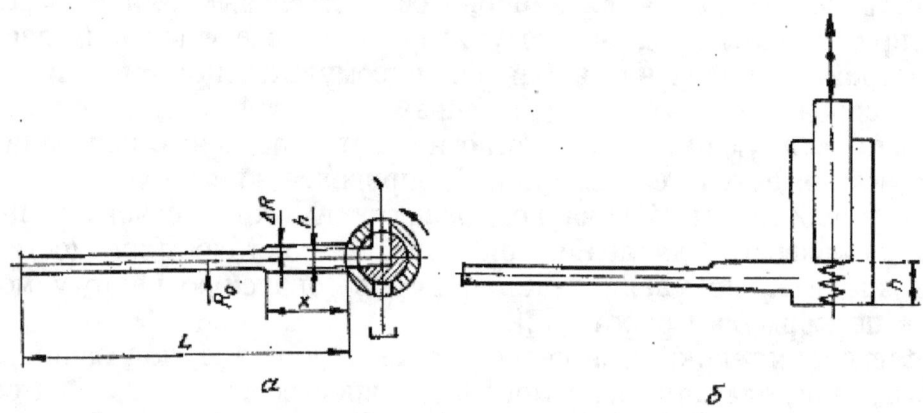

где R_0, δ, E, $\mu_{\text{п}}$, L — начальный внутренний радиус, толщина, модуль упругости, коэффициент Пуассона и длина трубопровода; t — время воздействия на него давления.

В случае применения трубчатых пружин в вибраторах крутильных колебаний угол поворота конца пружины под действием давления, центральный угол винтовой пружины и изменение объема внутренней полости ΔV (расход жидкости на один импульс) определяются из работы [1].

Работа источника давления за один импульс равна

$$A = p\Delta V.$$

При динамическом расчете гидроударного устройства уточняются конструктивные параметры, рабочие характеристики и проверяется устойчивость его работы. Сжимаемость жидкости не учитывается при $p > 2$ $Мн/м^2$ [6], волновые процессы — при длине трубопроводов, не соизмеримой с длиной волны, или частоте колебаний, не соизмеримой с периодом трубопровода (практически при $f < 50$ $гц$).

Уравнение расширения тупикового трубопровода под действием давления имеет вид

$$m (\Delta R)'' + \frac{4\pi E\delta L \Delta R}{R_0 (1 - 0,5\mu_{\text{п}})} + P - pF = 0. \qquad (1)$$

Здесь m — масса трубопровода и подвижных элементов нагрузки; P — постоянная составляющая рабочей нагрузки.

На основании волновых уравнений гидроудара [4] получаем

$$p = p_0 + \rho a [\varphi (x - at) + \psi (x + at)];$$

$$v - v_0 = \varphi (x - at) - \psi (x + at),$$

где x — координата по длине трубопровода; v и v_0 — текущая и начальная скорости жидкости в трубопроводе; $\varphi (x - at)$ и $\psi (x + at)$ — произвольные функции прямой и обратной волн, зависящие от граничных условий; ρ — плотность жидкости; a — скорость ударной волны.

Прямая волна, образующаяся в начале трубопровода ($x = 0$), распространяется по нему со скоростью a. Через время $t = \dfrac{L}{a}$ она достигнет конца трубопровода ($x = L$), в результате чего образуется отраженная обратная волна, распространяющаяся с той же скоростью в обратном направлении. Таким образом, значения функций прямой и обратной волн в любой момент времени равны соответственно значениям функций этих волн в граничных сечениях трубопровода

$$\varphi(x - at) = \varphi\left[-a\left(t - \frac{x}{a}\right)\right];$$

$$\psi(x + at) = \psi\left[L + a\left(t - \frac{L-x}{a}\right)\right].$$

Рассмотрим текущие давление p_i и скорость жидкости v_i через периоды времени $t = \dfrac{L}{a}$

$$p_i = p_0 + \rho a\,(\varphi_i + \psi_{i-1}); \quad v_i - v_0 = \varphi_i - \psi_{i-1} \quad (x = 0);$$

$$p_i = p_0\,(\varphi_{i-1} + \psi_i); \qquad v_i - v_0 = \varphi_{i-1} - \psi_i \quad (x = L). \tag{2}$$

Уравнение расхода гидроударных устройств при $x = L = 0$, когда длина трубопровода небольшая и волновые процессы отсутствуют, имеет вид:

тип a

$$2\pi R_0 L\,(\Delta R)'\,\dot{p}\,\frac{V_\text{ж}}{E_\text{ж}} - \mu b y \sqrt{2\,\frac{p}{\rho}} = 0; \tag{3}$$

тип $б$

$$2\pi R_0 L\,(\Delta R)' + \dot{p}\,\frac{V_\text{ж}}{E_\text{ж}} - Fwt = 0. \tag{4}$$

Здесь $E_\text{ж}$ — модуль упругости рабочей жидкости; μ, b, y — коэффициент расхода, ширина и текущая высота рабочего окна распределителя; $y = h\sin\omega_\text{р}t = h\omega_\text{р}t$; h — высота рабочего окна распределителя; $\omega_\text{р}$ — угловая скорость распределителя (частота колебаний клапанного пульсатора); $V_\text{ж}$ — объем жидкости в трубопроводе (так как деформация последнего невелика, считаем $V_* = \text{const}$); F, w — площадь и ускорение плунжера мультипликатора, на который воздействует ударный механизм.

Если $E_\text{ж} = \infty$ (жидкость несжимаема) и отсутствуют волновые процессы, то получаем:

тип a

$$m\,(\Delta R)'' - [(\Delta R)']^2\,a_1\cos ec^2\,\omega_\text{р}t + Z\Delta R = -P; \tag{5}$$

тип $б$

$$\Delta R = 0{,}08wt^2 F R_0^{-1} L^{-1}; \tag{6}$$

$$p = wF R_0^{-2}\left(0{,}025mL^{-2} + \frac{E\delta t^2}{1 - 0{,}5\mu_\text{п}}\right) + 0{,}64 P R_0^{-1} L^{-1}. \tag{7}$$

Здесь $a_1 = 122 R_0^3 L^3 \rho\,(\mu bh)^{-2}$; $Z = \dfrac{12{,}5 E\delta L}{R_0\,(1 - 0{,}5\mu_n)}$.

Для исследования устойчивости работы уравнение (5) линеаризуем методом кусочно-линейной аппроксимации [5]

$$m(\Delta R)'' - A_1 \Delta R_{\text{м}} a_1 \cosec^2 \omega_p t\, T^{-1}(\Delta R)' + Z\Delta R = -P -$$
$$- B\Delta R_{\text{м}}^2\, T^{-2} a_1 \cosec^2 \omega_p t, \tag{8}$$

где $A_1 = 0{,}38;\ B = 0$ при $0 \leqslant (\Delta R)' \leqslant 0{,}46\dfrac{\Delta R_{\text{м}}}{T}$;

$\qquad A_1 = 1{,}46;\ B = 0{,}5$ при $0{,}46\dfrac{\Delta R_{\text{м}}}{T} \leqslant (\Delta R)' \leqslant \dfrac{\Delta R_{\text{м}}}{T}$;

T — время полного перемещения стенки трубопровода.

Для работы системы в колебательном режиме в соответствии с критерием Гурвица необходимо, чтобы коэффициенты уравнения были отрицательными, что в данном случае обеспечивается.

При учете сжимаемости жидкости, но без учета волновых явлений, методом кусочно-линейной аппроксимации [5] получаем:

тип *а*

$$m(\Delta R)''' - A_2 a_2 m \sin \omega_p t (\Delta R)'' + \left(39{,}5R_0^2 L^2 \frac{E_{\text{ж}}}{V_{\text{ж}}} + Z\right)(\Delta R)' -$$
$$- A_2 a_2 Z \sin \omega_p t\, \Delta R = A_2 a_2 \sin \omega_p t\, (P + 6{,}28 R_0 L B_1 p_{\text{м}}); \tag{9}$$

тип *б*

$$m(\Delta R)''' + \left(39{,}5R_0 L^2 \frac{E_{\text{ж}}}{V_{\text{ж}}} + Z\right)(\Delta R)' = 6{,}28 R_0 L F E_{\text{ж}} V_{\text{ж}}^{-1} wt \cdot \tag{10}$$

Здесь

$$A_2 = 4{,}53;\ B_1 = 0 \quad \text{при } 0 \leqslant p \leqslant 0{,}07 p_{\text{м}};$$

$$A_2 = 0{,}78;\ B_1 = 0{,}27 \text{ при } 0{,}07 p_{\text{м}} \leqslant p \leqslant p_{\text{м}};$$

$$a_2 = \frac{1{,}41 E_{\text{ж}} \mu bh}{V_{\text{ж}} \sqrt{\rho p_{\text{м}}}}.$$

Анализ полученных уравнений показывает, что колебательный процесс обеспечивается конструкцией устройств. Следует заметить, что погрешность аппроксимации не превышает 4% максимального значения аппроксимируемой функции.

При учете волновых процессов и пренебрежении сжимаемостью жидкости определяем граничные условия для начала трубопровода:

тип *а*

$$\mu by \sqrt{2\frac{p_i}{\rho}} + \pi R_0^2 v_i = 0; \tag{11}$$

тип *б*

$$Fwt + \pi R_0^2 (v_i - v_0) = 0. \tag{12}$$

Совместное решение с уравнениями (2) дает:

тип *а*

$$\pi R_0^2 \left(2\psi_{i-1} - \frac{p_i - p_0}{\rho a}\right) - \mu by \sqrt{2\frac{p_i}{\rho}} = 0; \tag{13}$$

тип *б*

$$\pi R_0^2 \left(2\psi_{i-1} - \frac{p_i - p_0}{\rho a}\right) - Fwt = 0. \tag{14}$$

Зная функцию обратной волны, можно с помощью этих уравнений определить повышение давления в начале трубопровода для любого периода (например, графическим методом [3]).

Функция прямой волны для сечений $x = 0$ и $x = L$ равна

$$\varphi_i = \begin{cases} \dfrac{p_i - p_0}{\rho a} - \psi_{i-1} & (x = 0); \\[2mm] \varphi_{i-1} - (v_i - v_0) & (x = L). \end{cases} \quad (15)$$

Решая совместно уравнения (13), (14), (15) с учетом начальных условий (в первом периоде обратная волна отсутствует), можно получить изменение давления в трубопроводе через промежутки времени. Подстановкой найденных значений в уравнение (1) определяется перемещение стенки трубопровода в функции времени.

При $p < 2$ *Мн/м²* волновые процессы интенсивно затухают и не учитываются.

ЛИТЕРАТУРА

1. А н д р е е в а Л. Е., Упругие элементы приборов, М., Машгиз, 1962.
2. Б а ш т а Т. М., Машиностроительная гидравлика, М., Изд-во «Машиностроение», 1971.
3. Ж м у д ь А. Е., Гидравлический удар в гидротурбинных установках, М. — Л., Госэнергоиздат, 1953.
4. Ж у к о в с к и й Н. Е., О гидравлическом ударе в водопроводных трубках, М. — Л., Гостехиздат, 1949.
5. Ф и ш г а л С. И., Об аппроксимации функций сопротивления и расхода, Прикладная механика, т. VII, в. 9, 1971.
6. Ф и ш г а л С. И., Гидроударное определение парогазосодержания жидкости, Заводская лаборатория, № 12, 1972.

Поступила
17.IV 1972 г.

Киевский завод
«Стройдормаш»

IV.4. HYDRAULIC IMPACT NUT-RUNNERS
Machines & Tooling. Vol. XLI, London, 1970, No. 1
Production Engineering Research Association of Great Britain

MACHINES & TOOLING

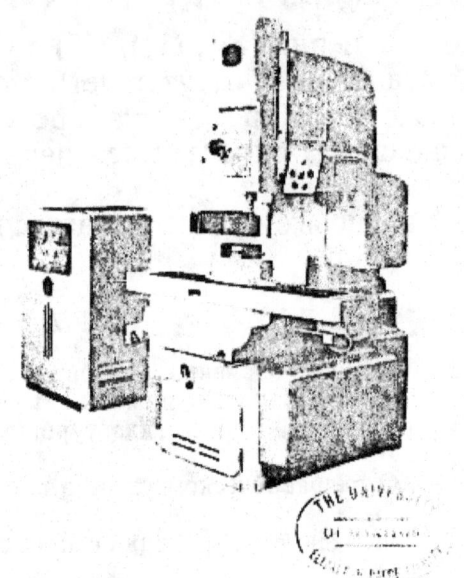

PRODUCTION ENGINEERING RESEARCH ASSOCIATION OF GREAT BRITAIN

ABSTRACT

Unlike known compressed air and hydraulic hut-runners, the below hydraulic impact (shock) ones have much higher torque/weight ratio, the step control possibility, no mechanically interacting components and therefore the low noise level.

HYDRAULIC IMPACT NUT-RUNNERS

S.I. FISHGAL

The main advantages of hydraulic impact-type nut-runners compared with conventional air-operated or hydraulic nut-runners are their high specific torque (ratio of the torque to the runner weight), absence of mechanical interaction of component parts, possibility of stepless torque control and low noise-level.

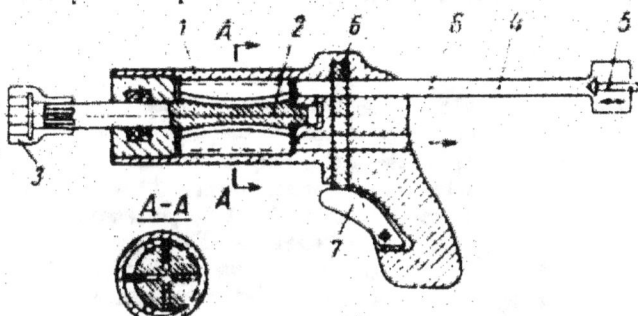

Fig. 1. *Hydraulic pulse nut-runner.*

The nut-runner depicted in Fig. 1 consists of body 1 in which is mounted vane rotor 2 of a hydraulic motor. The rotor is connected to a spindle which carries socket 3. Pressure oil is supplied via pipe 4, and input port and valve 6 (operated by trigger mechanism 7). After completing its work circuit the oil is drained through outlet port 8. The water-hammer generating mechanism is mounted in the pressure line but, in Fig. 1, is shown as valve 5; its function is to rapidly close input port 6 from the oil-supply pipe 4. When the input port is suddenly closed, water hammer is generated within the system, and is absorbed by the hydraulic motor. The hammer-generating valve may be also mounted downstream from the nut runner.

The performance, design and testing of this nut runner have been described previously[1],[2]. When a

vibratory pump is used as the pressure source then there is no need for a hammer-generating valve.

Another version of the nut runner[3] has a cylinder which receives the hammer pulses[3]. Between the reciprocating element 1 (see Fig. 2) and the driving

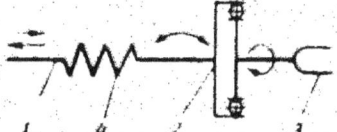

Fig. 2. *Nut runner with reciprocating motion (schematic)*

member of the free-wheel clutch 2 (with nut-runner socket 3) is located compression spring 4. The element 1 continuously compresses and releases the spring; when the spring is compressed it tends to unwind, and when it is released it tends to compress. Because the ends of the spring are rigidly secured during each compression and release they rotate in relation to each other. In this way the spring converts the reciprocating motion of the power element into oscillating motion of the driving free-wheel member. The free-wheel clutch 2 imparts to the runner socket pulse motions in one direction only. The angle of the relative rotation of the spring faces, and the axial deformation of the spring are determined using formulae given by Andreeva[4].

If a tubular-type spring is used it combines the functions of a power element and converter of linear oscillations into torsional oscillations. Such an arrangement of a nut runner is very simple; one end of the tubular spring is secured to the free-wheel clutch 2, carrying socket 3, and the other end is secured to the body 4. The oval-shaped bore of the spring, closed at one end, is connected to a source of pressure

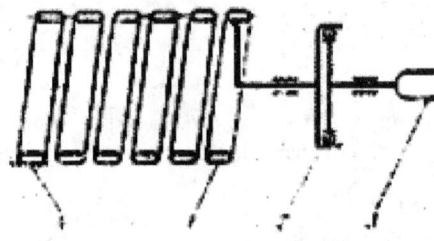

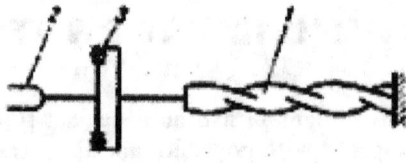

Fig. 4. *Nut runner with spiral tubular spring (schematic).*

Fig. 3. *Nut runner with cylindrical tubular spring (schematic).*

pulses (not shown). Fluid, under pressure, delivered to the spring chamber tends to deform it by making the cross-section circular. The longitudinal fibres, tending to retain their initial shape twist the spring thereby unwinding it. When the pressure is reduced the spring again compresses. Thus, the pressure pulses in the tubular spring are converted into torsional oscillations, which the free-wheel clutch converts into unidirectional runner pulses.

Depending on the design requirements the tubular spring may be replaced by a Bourdon-type spring, or spiral tubular spring. For a Bourdon-type spring the relative angle of rotation, caused by a pressure, is calculated from formulae by Andreeva[4]. The calculation procedures for spiral tubular springs, taking into account helix angles, have not yet been evolved but, considering the type of spiral used, these angles may be disregarded. The central angle, the angle of rotation of the spring end, changes in the spring volume (fluid consumption per pulse) of both types of spring, and the work done by the pressure source per pulse are also determined from the Andreeva formulae[4].

The schematic diagram in Fig. 4 shows a nut runner using a spiral tubular spring 1, with oval-shaped internal chamber. One end of the spring is closed, and is secured to a free-wheel clutch 3 with socket 2; the other end is secured to the body. The chamber is connected to a pressure-pulse source (not shown). Due to internal pressure the cross-section of the spring is

deformed, thereby altering the distance between the central axis of the section and the longitudinal fibres, causing a change in the spring length. Because the fibres tend to return to their initial length, the spring unwinds in relation to its axis. When the pressure is reduced the spring again compresses. Thus, pressure pulses in the spring chamber are converted into torsional oscillations, which the free-wheel clutch converts to uni-directional oscillations of the runner 2.

Nut runners with tubular springs are connected to a pump via a three-way rotary valve connecting the spring chamber alternately with the supply and drain lines[3]. Various types of hydraulic nut-runners for mechanizing assembly operations have been developed, by the GosavtodorNII and NItraktorsel'khozmash institutes, for torques from 4 to 80 mkg (the lower values apply to the nut runners depicted in Fig. 3 and 4).

REFERENCES

1 FISHGAL, S.I. *Hydraulic pulse drives of mechanical gears.* Stroitel'nye i dorozhnye mashiny, No. 10, 1968.

2. FISHGAL, S.I. *Investigation of hydraulic system elements subjected to pressure pulses.* UkrNIINTI. K., No. 6, 1968.

3. FISHGAL, S.I. *Nut-runner.* Author's Claim No. 254 407.

4. ANDREEVA, L.E. *Elastic elements of instruments.* Mashgiz. M., 1962.

IV.5. HYDRAULIC SHOCK NUT-RUNNERS
S.I. Fishgal. Machines & Tooling. Moscow, 1970, No. 1

ABSTRACT

In comparison with known compressed air and hydraulic hut-runners, the described hydraulic impact (shock) ones have much higher torque/weight ratio, the step control possibility, no mechanically interacting components and therefore the low noise level.

NOTE. Production Engineering Research Association of Great Britain translated this paper for its "Machines & Tooling." Vol. XLI, London, 1970, No. 1

УДК 621.883.3

С. И. Фишгал

Гидроударные гайковерты

Основные преимущества гидроударных гайковертов по сравнению с распространенными пневматическими и гидравлическими инструментами ударного действия заключаются в наличии значительного удельного крутящего момента (отношение крутящего момента к весу), отсутствии вредного механического взаимного воздействия деталей, возможности плавного изменения крутящего момента, а также в невысоком уровне шума.

Гидроударный гайковерт выполнен в виде корпуса *1* (рис. 1), в котором расположен ротор *2* лопастного гидродвигателя; ро-

тор соединен со шпинделем, несущим гаечный ключ *3*. Рабочая жидкость подается в гидродвигатель по трубопроводу *4* через входное отверстие и золотник *6* с приводом от курка *7*; сливается жидкость через отверстие *8*. На нагнетательной магистрали установлен механизм, создающий гидравлические удары; на рис. 1 этот механизм показан условно в виде клапана *5*, перекрывающего с большой скоростью проходное отверстие трубопровода *4*. При перекрытии трубопровода, по которому рабочая жидкость поступает из гидросистемы в гидродвигатель гайковерта, возникает гидравлический удар, воспринимаемый ротором. Клапан может быть установлен также и за сливной магистрали.

Характер работы гидроударного привода, элементы его расчета, испытание и конструкция пульсирующего устройства описаны в работах [1 и 2]. При использовании насоса вибрационного действия в качестве источника давления необходимость применения пульсатора отпадает.

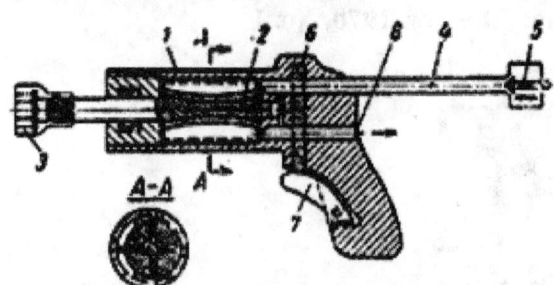

Рис. 1. Гидроударный гайковерт

Иначе выполнен гайковерт [3], в котором в качестве приемника колебаний применен цилиндр. Между возвратно-поступательно движущимся элементом *1* (рис. 2) силового органа и ведущей полумуфтой обгонной муфты *2* с ключом *3* жестко закреплена пружина *4* сжатия. Элемент *1* сообщает пружине возвратно-поступательное движение; при сжатии пружина стремится раскручиваться, а при растяжении — наоборот. Благодаря жесткому креплению торцов пружины при каждом ее сжатии-растяжении происходит их взаимный поворот. Таким образом, пружина преобразует возвратно-поступательное движение силового органа в крутильные колебания ведущей полумуфты. Обгонная муфта *3* сообщает ключу *3* одностороннее импульсное вращение. Взаимный угол поворота торцов пружины и величина ее осевой деформации определяются по формулам из работы [4].

Рис. 2. Схема гайковерта возвратно-поступательного действия

При применении трубчатой пружины последняя совмещает в себе функции силового органа и преобразователя продольных колебаний в крутильные. Конструкция такого гайковерта (рис. 3) очень проста. Конец цилиндрической пружины *1* прикреплен к обгонной муфте *2* с ключом *3* и к базовой детали *4*. Полость пружины овальной формы закрыта с одной стороны и присоединена к источнику пульсирующего давления (на рис. 3 не показан). Под давлением рабочей жидкости, подаваемой в эту полость, сечение пружины деформируется, приближаясь к форме окружности. Продольные волокна, стремясь сохранить свои первоначальные размеры, поворачивают поперечные сечения пружины, и последняя раскручивается; при снижении давления она закручивается. Таким образом, при подаче в полость пружины пульсирующего давления крутильные колебания сообщаются обгонной муфте *2*, преобразующей их в одностороннее импульсное вращение ключа *3*.

Вместо цилиндрической пружины в зависимости от требований конструкций могут быть применены пружина Бурдона или спиральная трубчатая пружина Бойса. Для пружины Бурдона относительный угол поворота конца пружины под дей-

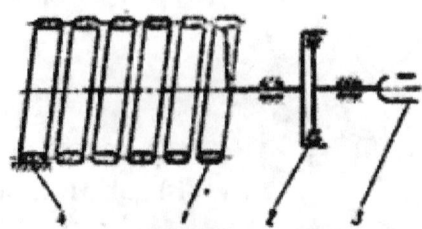

Рис. 3. Схема гидроударного гайковерта с цилиндрической трубчатой пружиной

ствием давления определяется из работы [4]. Расчет цилиндрических трубчатых пружин с учетом угла подъема винтовой линии еще не разработан; однако из-за малых значений последнего им можно пренебречь. Центральный угол, угол поворота конца и изменение объема внутренней полости (расход рабочей жидкости на один импульс) цилиндрической трубчатой пружины и пружины Бурдона, а также работа источника давления за один импульс определяются из работы [4].

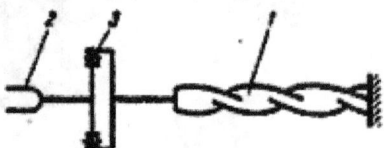

Рис. 4. Схема гидроударного гайковерта с витой трубчатой пружиной

На рис. 4 показана схема гидроударного гайковерта с витой трубчатой пружиной *1*, выполненной с овальным отверстием. Один конец пружины закрыт и присоединен к обгонной муфте *3* с ключом *2*, а другой к базовой детали. Полость пружины подсоединена к источнику пульсирующего давления (на рис. 4 не показан). Под действием внутреннего давления поперечное сечение пружины деформируется; при этом изменяются расстояния между центральной осью этого сечения и продольными волокнами, что приводит к изменению их длины. Благодаря стремлению волокон вернуться к первоначальной длине пружина раскручивается относительно оси; при снижении давления пружина закручивается. Таким образом, при подаче в полость пружины жидкости под пульсирующим давлением обгонной полумуфте сообщаются крутильные колебания, которые она преобразует в одностороннее импульсное вращение ключа *2*.

Описанные гайковерты с трубчатыми пружинами подсоединяют к насосной станции посредством вращающегося трехходового распределителя, соединяющего полости пружин поочередно с напорной и сливной магистралями [2]. Более проста установка для создания пульсирующего давления с пневмогидравлическим приводом [2]. Гидравлические гайковерты разных типов и модификаций для механизации механо-сборочных операций разработаны институтами «ГосавтодорНИИ» и «НИИтракторосельхозмаш» на крутящий момент от 4 до 80 кГм (меньшая величина относится к гайковертам на рис. 3 и 4).

Литература

1. Фишгал С. И. Гидроударный привод челюстей грейфера. «Строительные и дорожные машины», 1968, № 10.

2. Фишгал С. И. Испытание элементов гидросистем под пульсирующим давлением. Сб. «Технология и организация производства». К., УкрНИИНТИ, 1968, № 6.

3. Фишгал С. И. Гайковерт. Авторское свидетельство № 254 407.

4. Андреева Л. Е. Упругие элементы приборов. М., Машгиз, 1962.

IV.6. PERCUSSION GRAB-BUCKET SHELL
Construction and Road-Making Machines. Moscow, 1968, No. 10
(in Russian)

ABSTRACT

A hydraulic grab is provided with a power cylinder having a pressure cavity connected with a pump by a pipeline. For easier striking of the grab-bucket shell into a scooped material, an impulse exciter of hydraulic shocks is installed in the pipeline. The length of the latter is greater than a half product of the sound velocity and the impulse time.

The exciter is a rotary distributor transforming continuous flow into the pulsing one. It contains a turbine with a body, a rotor and vanes. The turbine is driven into rotation by the distributed flow. To simplify the structure the driving and distributing functions are combined together by executing the rotor floating and self-aligning in the cylindrical chamber of the body, and the vanes are hinged on the rotor's periphery.

Another vibration hydrodrive represents a vibration pump combining the exciter. The pump contains a body with three screw rotors. The middle one is drives the side screws. The latter covers a drainage orifice periodically. When the orifice is closed, the pump works as a static one.

№ 10, 1968 г.   Строительные и дорожные машины   29

УДК 621.879.062.3.001.2

Гидроударный привод челюстей грейфера

Инж. С. И. ФИШГАЛ

Вибрация челюстей грейфера является эффективным средством, уменьшающим сопротивление внедрению их в плотно связанный материал и увеличивающим набор его за один цикл. Поэтому представляет интерес конструкция гидравлического грейфера, в напорном цилиндре механизма замыкания челюстей которого создается пульсирующее давление.

Челюсти *1* (рис. 1) грейфера замыкаются при подаче рабочей жидкости под поршень *2* гидроцилиндра *3*, напорная полость которого соединена с гидросистемой базовой машины при помощи трубопровода *4* и механизма, создающего гидравлические удары. Пульсирующий клапан *5* с большой скоростью перекрывает проходное отверстие трубопровода *4*.

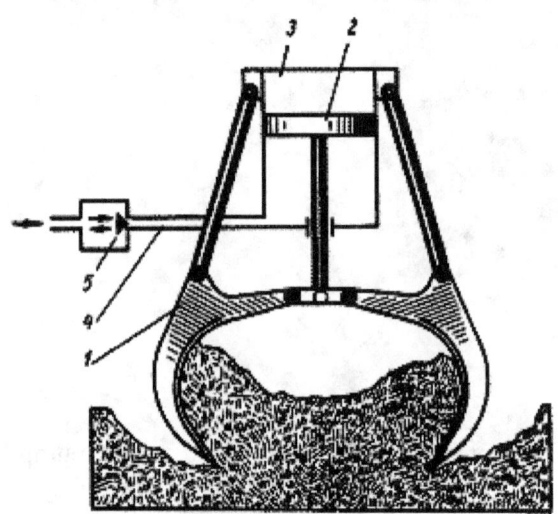

Рис. 1. Схема гидропривода челюстей грейфера

При перекрытии поршень *2* и рабочая жидкость продолжают еще по инерции перемещаться, замыкая челюсти. Жидкость при этом отрывается от клапана *5*. Затем под действием сил тяжести направление движения поршня *2* и жидкости меняется и последняя устремляется в обратном направлении к клапану *5*, который в это время закрыт, вследствие чего возникает гидравлический удар. Движение слоя жидкости, непосредственно соприкасающегося с клапаном, прекращается и кинетическая энергия этого слоя переходит в работу сжатия с мгновенным повышением давления, передающимся следующему слою, и т. д. Увеличение давления распространяется по трубопроводу *4* в виде волны со скоростью распространения звука в упругой среде. При достижении этой волной поршня *2* происходит встряхивание черпаемого материала и дальнейшее внедрение челюстей грейфера. Затем под действием сил тяжести направление движения поршня *2* и жидкости опять меняется и по трубопроводу *4* в направлении к клапану *5* распространяется волна разрежения, т. е. процесс повторяется, быстро, однако, затухая. Для его поддержания необходимо, чтобы клапан *5* открывался в момент отражения от него ударной волны. При этом он должен пропустить некоторое количество рабочей жидкости, необходимое для дальнейшего перемещения поршня *2*.

Количественный анализ волновых процессов в трубопроводе при подвижном поршне дан в работе [1].

В начале процесса, когда скорость поршня равна нулю, приращение давления в цилиндре в момент подхода к поршню переднего фронта прямой волны равно удвоенной величине превышения давления прямой волны над начальным давлением в трубопроводе. В дальнейшем под действием давления поршень приходит в движение. По мере возрастания скорости поршня снижение давления в цилиндре. Степень влияния скорости поршня на давление в цилиндре зависит от соотношения площадей поршня и трубопровода. Чем боль-

ше эта величина, тем быстрее происходит спад давления у поршня при отражении от него волны.

Время перекрытия клапана

$$t < \frac{2l}{c},$$

где *l* — длина трубопровода от клапана к цилиндру;

c — скорость распространения ударной волны в трубопроводе:

$$c = \frac{1}{\sqrt{\rho_0 \left(\frac{1}{E'} + \frac{d}{\delta} \cdot \frac{1}{E''} \right)}};$$

ρ_0 — плотность жидкости при атмосферном давлении;

E' — модуль упругости жидкости;

E'' — модуль упругости материала трубопровода;

d — диаметр трубопровода;

δ — толщина стенки трубопровода.

Максимальное давление у плунжера в момент подхода к нему ударной волны

$$p_п = 2p_з - p_0,$$

где $p_з$ — давление в гидросистеме;

p_0 — начальное давление в трубопроводе.

Следует учесть, что в трубопроводе от клапана *5* к насосу также возможно при определенных условиях возникновение гидравлических ударов. Происходит это следующим образом. При перекрытии слой жидкости, непосредственно соприкасающийся с клапаном на участке трубопровода от насоса к клапану, останавливается, и его кинетическая энергия переходит в работу сжатия с мгновенным повышением давления. Это повышение распространяется по участку трубопровода от клапана к насосу в виде волны сжатия. Так как давление жидкости в насосе в это время ниже, чем в трубопроводе, то часть жидкости из последнего устремляется в насос. После этого давление в трубопроводе снижается и становится меньше давления в насосе. Вследствие этого жидкость из насоса поступает обратно в трубопровод, повышая давление в последнем. Затем процесс повторяется.

Таким образом, при перекрытии трубопровода к напорной полости гидроцилиндра гидравлические удары могут возникать как на участке трубопровода от клапана к цилиндру, так и на участке от насоса к клапану. Последнее является, конечно, нежелательным.

Для устранения гидравлических ударов в участке трубопровода от насоса к клапану длину этого участка следует принять по возможности минимальной, а насосную установку желательно оборудовать гидравлическим аккумулятором. Кроме того, можно подавать жидкость в клапан через лабиринтные каналы, оказывающие сравнительно небольшое сопротивление потоку жидкости и весьма большое сопротивление распространению ударной волны.

Конструкция пульсирующего устройства[1] схематично изо-

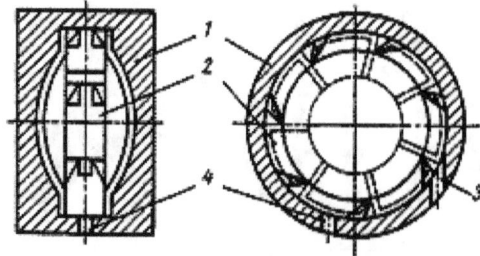

Рис. 2. Схема пульсирующего устройства

―――――――――
[1] Авторское свидетельство № 189264. Опубликовано в «Бюллетене изобретений и товарных знаков», 1966, № 23.

бражена на рис. 2. В корпус *1* помещен плавающий самоустанавливающийся ротор *2* с пустотелыми скошенными лопатками. Рабочая жидкость из гидросистемы поступает через входное отверстие *3* в корпус *1*, вращая ротор *2*. Под действием центробежных сил лопатки ротора *2* прижимаются к расточке корпуса *1* и перекрывают впускное *3* и выпускное *4* отверстия, преобразовывая тем самым непрерывный поток жидкости в пульсирующий. Так как лопатки ротора *2* выполнены скошенными, то жидкость распределяется по обе стороны ротора *2*, исключая его осевое смещение. Действительно, если по каким-либо причинам ротор смещен в одну сторону, то зазор между ротором и расточкой корпуса *1* в этом месте будет меньше, чем в противоположном. Вследствие этого здесь образуется повышение давления, под действием которого ротор *2* смещается в центральное положение.

При применении насоса вибрационного действия в качестве источника давления гидросистемы базовой машины отпадает необходимость использования пульсатора.

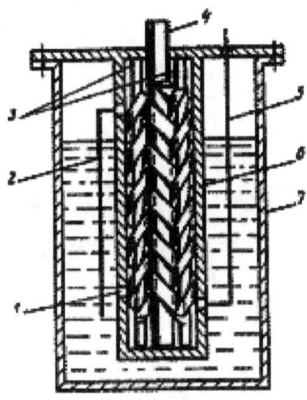

Рис. 3. Схема насоса

Насос (рис. 3) состоит из корпуса *6* с тремя винтовыми роторами, средний из которых является ведущим *4*, а два боковых — ведомыми *8*. Передаточное отношение между роторами *3* и *4* равно единице. Винты обычно двухзаходные, профиль — циклоидальный. Насос помещен в масляный бак *7*. Масло поступает в насос через всасывающий патрубок *2*, а выходит через напорный патрубок *5*. Полость высокого давления насоса соединяется непосредственно с жидкостью бака *7* через сбросное отверстие *1*.

Конструктивные соотношения винтов следующие [2]:

$$D_в = d_к; \quad D_н = \frac{5}{3} d_к; \quad d_в = \frac{1}{3} d_к;$$

$$t = \frac{10}{3} d_к; \quad L \approx 1{,}25\, t,$$

где $D_в$ — внутренний диаметр ведущего винта;
$D_н$ — наружный диаметр ведущего винта;
$d_в$ — внутренний диаметр ведомого винта;
$d_к$ — основной диаметр винтов; или наружный диаметр ведомого винта;
t — шаг винтов;
L — минимальная длина винтов из условия обеспечения герметичности.

При работе насоса ведомый винт гребнями периодически перекрывает сбросное отверстие *1*. Когда отверстие перекрыто, вибронасос работает как обычный статический насос с производительностью [2]

$$Q = \frac{d_к^3\, n\, \eta}{14{,}5 \cdot 10^3},$$

где n — число оборотов винта в минуту;
η — объемный к.п.д. насоса.

При полном открытии сбросного отверстия *1* жидкость перетекает из напорной полости во всасывающую с расходом

$$q = 0{,}785\, \mu\, d^2 \sqrt{\frac{2\Delta p}{\rho}},$$

где d — диаметр отверстия;
μ — коэффициент расхода отверстия;
Δp — разность давлений между полостями насоса;
ρ — плотность жидкости.

Таким образом, подача описываемого насоса пульсирующая. Для возникновения гидравлических ударов в напорном патрубке в соответствии с теорией гидроудара необходимо, чтобы время перекрытия сбросного отверстия *1* ротором *3* было меньше удвоенного отношения длины патрубка к скорости волны.

Частота колебаний давления в системе обусловливается частотой перекрытий отверстия *1* и для двухзаходного винта составляет $f = \dfrac{n}{30}$.

Литература

1. Тарко Л. М. Волновые процессы в трубопроводах гидромеханизмов. М., Машгиз, 1963.
2. Башта Т. М. Машиностроительная гидравлика. М., Машгиз, 1963.

IV.7. HYDRAULIC GRAB
Soviet Invention No. 449,870; 1964, by S.I. Fishgal

ABSTRACT

A hydraulic grab is provided with a power cylinder having a pressure cavity connected with a pump by a pipeline. For easier striking of the grab-bucket shell into a scooped material, an impulse exciter of hydraulic shocks is installed in the pipeline. The length of the latter is greater than a half product of the sound velocity and the impulse time.

Союз Советских
Социалистических
Республик

ОПИСАНИЕ ИЗОБРЕТЕНИЯ

К АВТОРСКОМУ СВИДЕТЕЛЬСТВУ

(11) **449870**

(61) Зависимое от авт. свидетельства —

(22) Заявлено 23.09.64 (21) 922070/27-11

с присоединением заявки № —

(32) Приоритет —

Государственный комитет
Совета Министров СССР
по делам изобретений
и открытий

Опубликовано 15.11.74. Бюллетень № 42

Дата опубликования описания 01.04.75

(51) М. Кл. B 66c 3/16

(53) УДК 621.86.063.25
(088.8)

(72) Автор
изобретения

С. И. Фишгал

(71) Заявитель

—

(54) ГИДРАВЛИЧЕСКИЙ ГРЕЙФЕР

1

Изобретение относится к подъемно-транспортной технике и может быть использовано при перегрузке сыпучих и крупно-кусковых материалов грейферными механизмами.

Известны грейферы для сыпучих материалов, содержащие корпус с насосом, силовым гидроцилиндром и шарнирно закрепленными на корпусе челюстями. Эти грейферы характеризуются небольшим заглублением в зачерпываемый материал и тем самым неполным использованием емкости грейфера.

Предлагаемый грейфер отличается тем, что в трубопроводе, соединяющем насос с напорной полостью силового гидроцилиндра, установлен возбудитель импульсных гидравлических ударов. Это облегчает внедрение челюстей в зачерпываемый материал. При этом длина трубопровода больше половины величины произведения скорости звука в рабочей среде на время импульса.

На чертеже представлено предлагаемое устройство.

Челюсти 1 грейфера открываются и закрываются посредством силового гидроцилиндра 2, имеющего напорную полость 3, соединенную трубопроводом 4 с насосом (на чертеже не показан). В трубопроводе 4 установлен возбудитель 5 импульсных гидравлических ударов. Трубопровод имеет длину не менее

2

половины величины произведения скорости звука в рабочей среде на время импульса.

В процессе работы раскрытый грейфер опускается на зачерпываемый материал, в напорной полости 3 силового гидроцилиндра 2 насосом создается давление, под действием которого челюсти 1 начинают закрываться. При этом возбудитель 5 импульсных гидравлических ударов систематически обеспечивает мгновенное перекрытие трубопровода 4, что сопровождается мгновенным повышением давления, которое, после открытия трубопровода, распространяется в виде волны по нему и напорной полости 3 со скоростью распространения звука в рабочей среде. Если длина трубопровода больше половины величины произведения скорости звука в рабочей среде на время импульса, то к моменту столкновения гидравлической волны с поршнем гидроцилиндра 2 она удваивает свое давление, создавая гидравлический удар, который передается на челюсти 1, облегчая их внедрение в зачерпываемый материал.

Предмет изобретения

1. Гидравлический грейфер, снабженный силовым гидроцилиндром, имеющим напорную полость, соединенную посредством трубопровода с насосом, отличающийся тем, что, с целью облегчения внедрения челюстей в

449870

3

зачерпываемый материал, в трубопроводе
установлен возбудитель импульсных гидравлических ударов.

2. Гидравлический грейфер по по. 1, отли-

4

ча ю щ и й с я тем, что длина трубопровода
больше половины величины произведения скорости звука в рабочей жидкости на время
импульса.

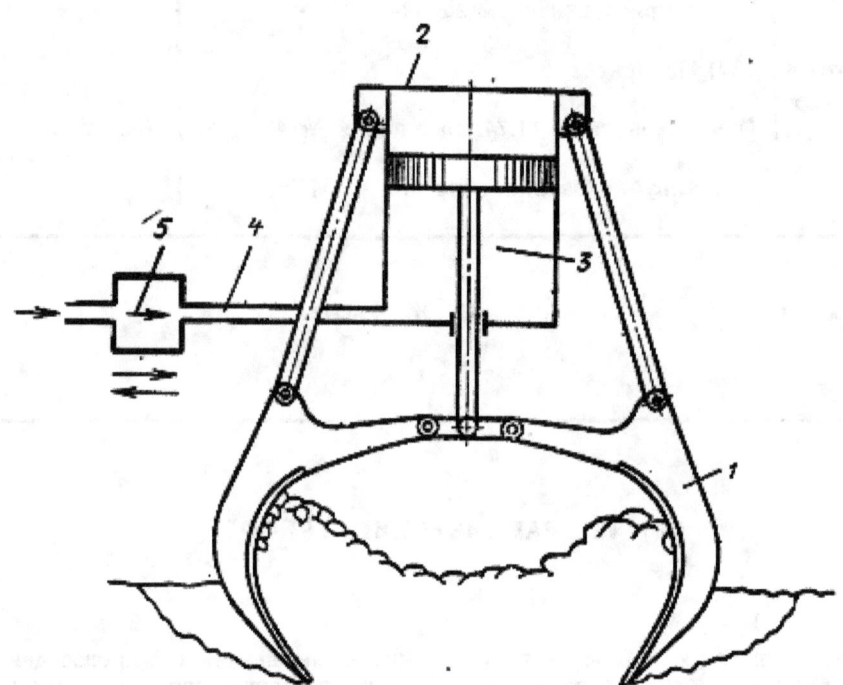

Составитель Б. Стрекалов

Редактор Л. Тюрина   Техред Г. Дворина   Корректор Н. Учакина

Заказ 743/4   Изд. № 1141   Тираж 811   Подписное
ЦНИИПИ Государственного комитета Совета Министров СССР
по делам изобретений и открытий
Москва, Ж-35, Раушская наб., д. 4/5

Типография, пр. Сапунова, 2

IV.8. METALS PLASMA ARC-CUTTING TORCH
Soviet Invention No. 195,299; 1965, by S.I. Fishgal

ABSTRACT

A metals plasma arc-cutting torch contains a chamber, an electrode longitudinal feed-off mechanism and systems for water cooling and forming a plasma jet. To provide repeating discharges between the electrode and the cut metal, the torch is connected to a reservoir from which electrolyte is injected into the discharge space between the electrode and the metal.

Союз Советских
Социалистических
Республик

Комитет по делам
изобретений и открытий
при Совете Министров
СССР

ОПИСАНИЕ ИЗОБРЕТЕНИЯ

К АВТОРСКОМУ СВИДЕТЕЛЬСТВУ

195299

Зависимое от авт. свидетельства № —

Заявлено 18.XI.1965 (№ 1037798/25-27)

с присоединением заявки № —

Приоритет —

Опубликовано 12.IV.1967. Бюллетень № 9

Дата опубликования описания 20.VI.1967

Кл. 49h, 35/02

МПК В 23k

УДК 621.791.945.55.03
(088.8)

Автор
изобретения   С. И. Фишгал

Заявитель   —

ГОРЕЛКА ДЛЯ ПЛАЗМЕННОЙ РЕЗКИ МЕТАЛЛОВ

Известна горелка для плазменной резки металлов, содержащая камеру с рабочим электродом, механизм для его продольной подачи и системы водяного охлаждения и формирования плазменной струи.

Предлагаемая горелка обеспечивает многократное образование разряда между рабочим электродом и разрезаемой деталью благодаря тому, что к горелке присоединен резервуар, из которого в разрядный промежуток между электродом и деталью поступает электролит.

На чертеже изображена описываемая горелка.

Горелка состоит из механизма 1 для продольной подачи электрода 2, электродной 3 и плазменной 4 камер, расположенных последовательно одна за другой. Электродная камера снабжена промежуточным соплом 5, а плазменная камера — выходным соплом 6. Система охлаждения горелки включает продольные 7 и винтовые 8 каналы для охлаждающей воды, система формирования плазменного пучка — тангенциальные каналы 9. По каналам 9 вода поступает в плазменную камеру, приобретая в последней окружную скорость для сжатия плазменного шнура. К горелке присоединен резервуар 10 с плавающим подпружиненным поршнем для создания давления. Резервуар заполнен электролитом.

Электролит и вода поступают через клапаны или краны 11, 12, управлять которыми можно одной рукояткой (механическая связь рукояток кранов изображена на чертеже условно). На горелку подают «минус» напряжения, на разрезаемую деталь 13 — «плюс».

Для разрезания детали следует на короткое время перевести краны в положение, изображенное на чертеже. При этом сливная и подводящая водопроводная магистрали перекрыты, а электролит поступает в горелку из резервуара, соединяя электрически электрод с разрезаемой деталью. Таким образом, струя раствора электролита является проводящим мостиком, при испарении электролита возникает стабильная дуга. Потом устанавливают краны в положение, при котором резервуар перекрыт, а горелка соединяется с водопроводной и сливной магистралями. При входе в горелку вода сначала попадает в электродную камеру и охлаждает ее. Затем поток воды разветвляется, часть ее по каналам 8 поступает в канал 7, где охлаждает электрод и участвует в образовании плазмы в электродном пространстве, часть воды впрыскивается тонкими струйками в пространство между соплами в нижней части горелки. Здесь благодаря тангенциальному расположению каналов 9 возникает вихревое водяное кольцо, сжимающее плазменный шнур, попадающий

195299

3

сюда через промежуточное сопло 5. Сливается вода из горелки через кран 11.

Для уменьшения расхода воды можно питать горелку водой по замкнутому циклу, используя общеизвестные средства (гидронасос, резервуар и др.).

Предмет изобретения

Горелка для плазменной резки металлов, содержащая камеру с рабочим электродом, 10

4

механизм для его продольной подачи и системы водяного охлаждения и формирования плазменной струи, отличающаяся тем, что, с целью обеспечения многократного образования разряда между рабочим электродом и разрезаемой деталью, к горелке присоединен резервуар, из которого в разрядный промежуток между электродом и деталью поступает электролит.

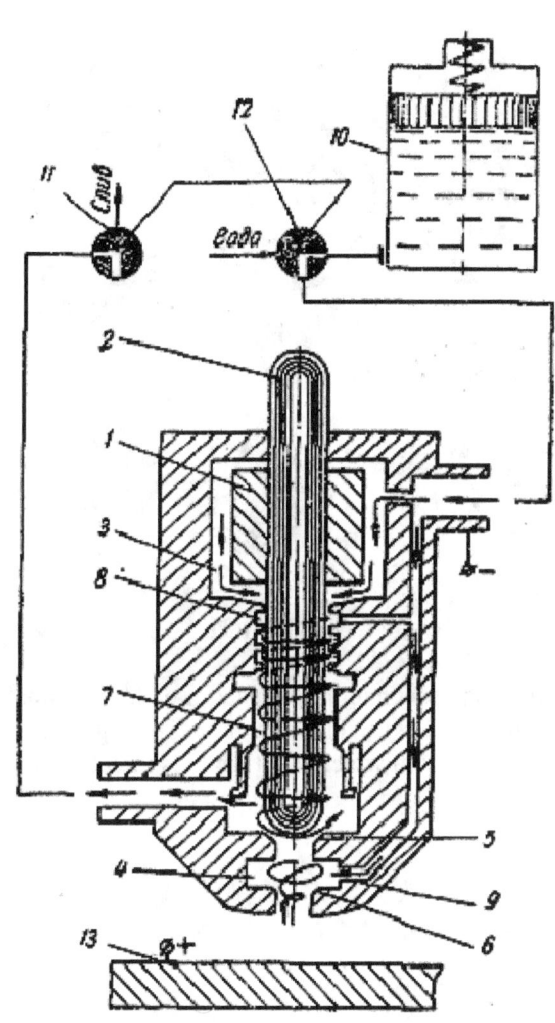

Составитель Т. Олесова

Редактор Л. Жаворонкова   Техред Т. П. Курилко   Корректоры: Л. В. Наделяева
                                                                    и В. В. Крылова

Заказ 1911/1   Тираж 535   Подписное
ЦНИИПИ Комитета по делам изобретений и открытий при Совете Министров СССР
                    Москва, Центр, пр. Серова, д. 4

Типография, пр. Сапунова 2

IV.9. LIQUID THIN JET - ELECTRODE!
S.I. Fishgal. Inventor and Rationalizer/Innovator,
Moscow, 1967, No. 8 (in Russian)

Изобретатель и Рационализатор

8 1967

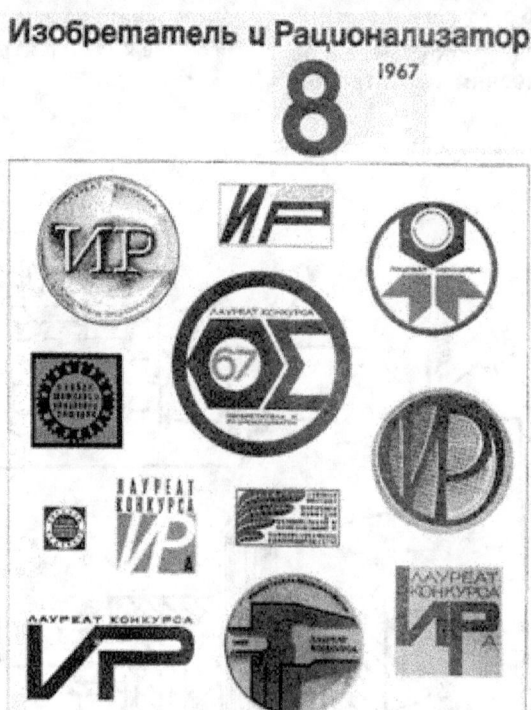

ABSTRACT

A metals plasma arc-cutting torch contains a chamber, an electrode longitudinal feed-off mechanism and systems for water cooling and forming a plasma jet. To provide repeating discharges between the electrode and the cut metal, the torch is connected to a reservoir from which electrolyte is injected into the discharge space between the electrode and the metal.

Изобретено в СССР

Струйка жидкости— электрод!

В плазменной горелке новой конструкции запалом плазменной дуги служит не алюминиевая проволока, которую каждый раз надо менять, а струйка электролита.

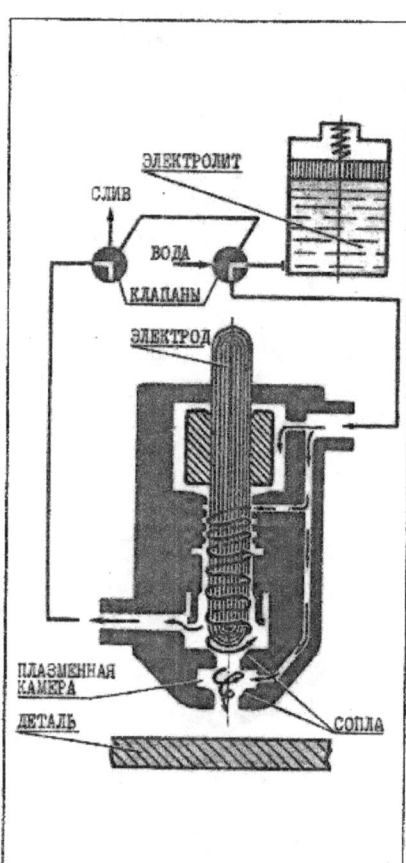

ЭЛЕКТРОЛИТ
СЛИВ
ВОДА
КЛАПАНЫ
ЭЛЕКТРОД
ПЛАЗМЕННАЯ КАМЕРА
ДЕТАЛЬ
СОПЛА

Цветные металлы, чугун и дюралюминий не подвластны ни автогену, ни электроду. Плазменные горелки превосходно режут любые металлы, но потребляют аргон или смесь его с водородом. Кроме «изысканного» питания, такой горелке нужен также тщательно центрируемый вольфрамовый электрод.

Горелка по чехословацкому патенту № 100498 (см. ИР, 4, 1965) не столь требовательна. В ее корпус подается вода, охлаждающая угольный (а не вольфрамовый!) электрод и цилиндрическую плазменную камеру, через которую проходит дуга.

В плазменную камеру вода подается по тангенциальным каналам. Благодаря этому поток воды завихряется, сжимая плазменный шнур. Температура при этом повышается почти до 20 тысяч градусов. Чехословацкая горелка легко режет и чугун, и цветные металлы, однако она имеет существенный недостаток. Чтобы получить дугу, сначала нужно коснуться детали электродом, а затем сразу же его отвести. Электрики об этом говорят так: следует образовать первичный токопроводящий мостик. В плазменной горелке это сделать непросто, так как мешает плазменная камера. Поэтому прежде чем установить в горелку электрод, в его торце сверлят отверстие, куда вставляют кусочек алюминиевой проволоки — своего рода «запал». Прикоснувшись к детали, проволока тут же сгорает, образуя дугу. Таким образом, для каждого нового реза нужно вынимать электрод, сверлить отверстие, вставлять проволоку и т. д. Резов же за смену приходится делать не один десяток.

Автор этих строк изобрел (а. с. № 195299) горелку, в которой разряд между электродом и разрезаемой деталью перебрасывается струей электролита.

Горелка состоит из механизма продольной подачи электрода, электродной и плазменной камер. Эти камеры последовательно расположены друг за другом. Каждая имеет по соплу. Горелка охлаждается водой, поступающей по продольным и винтовым каналам. В плазменную камеру вода попадает по тангенциальным каналам. Здесь поток приобретает окружную скорость и сжимает плазменный шнур.

Электролит, например соленая вода, помещается в специальном резервуаре. Давление в нем создается подпружиненным плавающим поршнем. Управляют подачей электролита два клапана, которые приводятся в действие одной рукояткой. На горелку подают «минус» напряжения, на разрезаемую деталь — «плюс».

При резке клапаны на мгновение поворачивают так, что сливная и подводящая водопроводная магистрали оказываются перекрытыми, а электролит из резервуара свободно поступает в горелку и электрически соединяет электрод с деталью. Таким образом, струя электролита служит первичным проводящим мостиком — «запалом». Она существует недолго и сразу же испаряется. После этого возникает стабильная дуга.

Затем клапаны ставят в такое положение, чтобы резервуар с электролитом был перекрыт, а горелка соединялась с водопроводной и сливной магистралями. Вода сначала попадает в электродную камеру и охлаждает ее, затем поток разветвляется. Часть его охлаждает электрод и участвует в образовании плазмы в электродном пространстве. Другая часть поступает в пространство между соплами в нижней части горелки. Здесь благодаря тангенциальному расположению каналов образуется вихревое водяное кольцо, сжимающее плазменный шнур.

Обычно вода из горелки уходит на слив. При питании горелки водой по замкнутому циклу расход ее значительно уменьшается.

С. ФИШГАЛ,
инженер

г. Киев

www.ingramcontent.com/pod-product-compliance
Lightning Source LLC
Chambersburg PA
CBHW081240180526
45171CB00005B/491